Solutions Manual
to Accompany
Accelerated Chemistry

John D. Mays

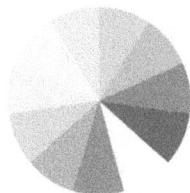

CENTRIPETAL PRESS
CLASSICAL ACADEMIC PRESS

Camp Hill, Pennsylvania
2021

CENTRIPETAL PRESS

CLASSICAL ACADEMIC PRESS

Solutions Manual to Accompany Accelerated Chemistry

© Classical Academic Press®, 2021

Edition 2.0

ISBN: 978-0-9981699-2-7

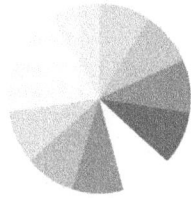

Classical Academic Press

515 S. 32nd Street

Camp Hill, PA 17011

www.ClassicalAcademicPress.com/Novare/

IS.04.23

Contents

Preface iii

Chapter 1 1

Chapter 2 14

Chapter 3 16

Chapter 4 18

Chapter 5 21

Chapter 6 29

Chapter 7 35

Chapter 8 57

Chapter 9 68

Chapter 10 83

Chapter 11 96

Chapter 12 113

Chapter 13 123

Appendix A 129

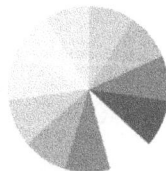

CENTRIPETAL PRESS

CLASSICAL ACADEMIC PRESS

Preface

This solutions manual contains fully detailed solutions for the computational problems contained in my text *Accelerated Chemistry*. Teachers and students using that text should find this manual to be a valuable resource.

Explanations for the number of significant digits shown in results are included for representative problems. Throughout this manual, significant digits are referred to as "sig digs."

When comparing your results to the results shown here and to those in the text, keep in mind that the last digit is always uncertain because of the way significant digits in measurements are defined. When two results match except for a difference of 1 in the most precise digit, we say that the results match. Because of rounding in calculators, it is not uncommon for results shown here to differ from the answer key in the text or from your result by 1 in the most precise digit.

I have checked and double checked the solutions to make them as accurate as possible. However, in any manual of this kind it is inevitable that errors remain. If you find an error, we would be much obliged if you would inform us of it by sending an email to info@centripetalpress.com.

Chapter 1

1.

$$\lambda = 543 \text{ nm} \cdot \frac{1 \text{ m}}{1 \times 10^9 \text{ nm}} = 5.43 \times 10^{-7} \text{ m}$$

$$E = \frac{hv}{\lambda} = \frac{6.626 \times 10^{-34} \text{ J} \cdot \text{s} \cdot 2.9979 \times 10^8 \frac{\text{m}}{\text{s}}}{5.43 \times 10^{-7} \text{ m}} = 3.66 \times 10^{-19} \text{ J}$$

2.

$$E = 2.2718 \times 10^{-19} \text{ J}$$

With 5 sig digs, we need to use constants from Appendix B with at least 5 sig digs:

$$E = \frac{hv}{\lambda} \rightarrow \lambda = \frac{hv}{E} = \frac{6.626 \times 10^{-34} \text{ J} \cdot \text{s} \cdot 2.9979 \times 10^8 \frac{\text{m}}{\text{s}}}{2.2718 \times 10^{-19} \text{ J}} = 8.7439 \times 10^{-7} \text{ m} \cdot \frac{10^9 \text{ nm}}{\text{m}} = 874.39 \text{ nm}$$

The result has 5 sig digs because the given value has 5.

3.

From Figure 1.6, $E = 2.18 \times 10^{-18}$ J

$$E = \frac{hv}{\lambda} \rightarrow \lambda = \frac{hv}{E} = \frac{6.626 \times 10^{-34} \text{ J} \cdot \text{s} \cdot 2.9979 \times 10^8 \frac{\text{m}}{\text{s}}}{2.18 \times 10^{-18} \text{ J}} = 9.11 \times 10^{-8} \text{ m} \cdot \frac{10^9 \text{ nm}}{1 \text{ m}} = 91.1 \text{ nm}$$

The result has 3 sig digs because the given value has 3.

4.

The visible spectrum is the Balmer series. From Figure 1.8,

$$\lambda_1 = 410 \text{ nm} \cdot \frac{1 \text{ m}}{1 \times 10^9 \text{ nm}} = 4.1 \times 10^{-7} \text{ m}$$

$$\lambda_2 = 434 \text{ nm} \cdot \frac{1 \text{ m}}{1 \times 10^9 \text{ nm}} = 4.34 \times 10^{-7} \text{ m}$$

$$\lambda_3 = 486 \text{ nm} \cdot \frac{1 \text{ m}}{1 \times 10^9 \text{ nm}} = 4.86 \times 10^{-7} \text{ m}$$

$$\lambda_4 = 656 \text{ nm} \cdot \frac{1 \text{ m}}{1 \times 10^9 \text{ nm}} = 6.56 \times 10^{-7} \text{ m}$$

$$E_1 = \frac{hv}{\lambda} = \frac{6.626 \times 10^{-34} \text{ J} \cdot \text{s} \cdot 2.9979 \times 10^8 \frac{m}{s}}{4.1 \times 10^{-7} \text{ m}} = 4.8 \times 10^{-19} \text{ J} \left(2 \text{ sig digs because 410 has 2}\right)$$

$$E_2 = \frac{hv}{\lambda} = \frac{6.626 \times 10^{-34} \text{ J} \cdot \text{s} \cdot 2.9979 \times 10^8 \frac{m}{s}}{4.34 \times 10^{-7} \text{ m}} = 4.58 \times 10^{-19} \text{ J} \left(3 \text{ sig digs because 434 has 3}\right)$$

$$E_3 = \frac{hv}{\lambda} = \frac{6.626 \times 10^{-34} \text{ J} \cdot \text{s} \cdot 2.9979 \times 10^8 \frac{m}{s}}{4.86 \times 10^{-7} \text{ m}} = 4.09 \times 10^{-19} \text{ J} \left(3 \text{ sig digs because 486 has 3}\right)$$

$$E_4 = \frac{hv}{\lambda} = \frac{6.626 \times 10^{-34} \text{ J} \cdot \text{s} \cdot 2.9979 \times 10^8 \frac{m}{s}}{6.56 \times 10^{-7} \text{ m}} = 3.03 \times 10^{-19} \text{ J} \left(3 \text{ sig digs because 656 has 3}\right)$$

18.

In Table 1.5, the heaviest nuclide listed is U-238, which is $Z = 92$. 238 nucleons – 92 protons = 146 neutrons. The lightest nuclide listed is H-1, which is $Z = 1$. 1 nucleon – 1 proton = 0 neutrons.

---

19. a.

$$73.2 \text{ g Cu} \cdot \frac{1 \text{ mol}}{63.55 \text{ g}} \cdot \frac{6.022 \times 10^{23} \text{ particles}}{\text{mol}} = 6.94 \times 10^{23} \text{ particles (atoms)}$$

---

19. b.

$$1.35 \text{ mol Na} \cdot \frac{6.022 \times 10^{23} \text{ particles}}{\text{mol}} = 8.13 \times 10^{23} \text{ particles (atoms)}$$

---

19. c.

$$1.5000 \text{ kg W} \cdot \frac{1000 \text{ g}}{1 \text{ kg}} \cdot \frac{\text{mol}}{183.85 \text{ g}} \cdot \frac{6.0221 \times 10^{23} \text{ particles}}{\text{mol}} = 4.9133 \times 10^{24} \text{ particles (atoms)}$$

---

20. a

$$6.022 \times 10^{23} \text{ atoms K} \cdot \frac{\text{mol}}{6.022 \times 10^{23} \text{ atoms}} \cdot \frac{39.098 \text{ g}}{\text{mol}} = 39.10 \text{ g}$$

(4 sig digs because the given value has 4)

---

20. b.

$$100 \text{ atoms Au} \cdot \frac{\text{mol}}{6.022 \times 10^{23} \text{ atoms}} \cdot \frac{196.9665 \text{ g}}{\text{mol}} = 3 \times 10^{-20} \text{ g}$$

(1 sig dig because the given value has 1)

---

20. c.

$$0.00100 \text{ mol Xe} \cdot \frac{131.29 \text{ g}}{\text{mol}} = 0.131 \text{ g}$$

(3 sig digs because the given value has 3)

---

20. d.

$$2.0 \text{ mol Li} \cdot \frac{6.941 \text{ g}}{\text{mol}} = 14 \text{ g}$$

(2 sig digs because the given value has 2)

---

20. e.

$$4.2120 \text{ mol Br} \cdot \frac{79.904 \text{ g}}{\text{mol}} = 336.56 \text{ g}$$

(5 sig digs because the given value has 5)

---

20. f.

$$7.422 \times 10^{22} \text{ atoms Pt} \cdot \frac{\text{mol}}{6.022 \times 10^{23} \text{ atoms}} \cdot \frac{195.08 \text{ g}}{\text{mol}} = 24.04 \text{ g}$$

(4 sig digs because the given value has 4)

---

21. a.

$$25 \text{ g Ca(OH)}_2 \cdot \frac{\text{mol}}{74.09 \text{ g}} = 0.34 \text{ mol}$$

---

21. b.

$$286.25 \text{ g Al}_2(\text{CrO}_4)_3 \cdot \frac{\text{mol}}{401.944 \text{ g}} = 0.71216 \text{ mol}$$

---

21. c

$$2.111 \text{ kg KCl} \cdot \frac{1000 \text{ g}}{1 \text{ kg}} \cdot \frac{\text{mol}}{74.551 \text{ g}} = 28.32 \text{ mol}$$

---

21. d.

$$47.50 \text{ g LiClO}_3 \cdot \frac{\text{mol}}{90.39 \text{ g}} = 0.5255 \text{ mol}$$

---

21. e.

$$10.0 \text{ g O}_2 \cdot \frac{\text{mol}}{32.0 \text{ g}} = 0.313 \text{ mol}$$

---

21. f.

$$1.00 \text{ mg C}_{14}\text{H}_{18}\text{N}_2\text{O}_5 \cdot \frac{1 \text{ g}}{1000 \text{ mg}} \cdot \frac{\text{mol}}{294.307 \text{ g}} = 3.40 \times 10^{-6} \text{ mol}$$

22.

Using data from Tables 1.5 and 1.6:

92 protons $92 \cdot 1.007276 \text{ u} = 92.66939 \text{ u}$

146 neutrons $146 \cdot 1.008665 \text{ u} = 147.2651 \text{ u}$

92 electrons $92 \cdot 0.0005486 \text{ u} = 0.05047 \text{ u}$

$$
\begin{array}{r}
92.66939 \text{ u} \\
147.2651 \text{ u} \\
+ \quad 0.05047 \text{ u} \\
\hline
239.9850 \text{ u}
\end{array}
$$

$$
\begin{array}{r}
239.9850 \text{ u} \\
- \quad 238.0508 \text{ u} \\
\hline
1.9342 \text{ u}
\end{array}
$$

$$1.9342 \text{ u} = 1.9342 \ \frac{\text{g}}{\text{mol}} \cdot \frac{\text{mol}}{6.02214076 \times 10^{23} \text{ atoms}} = 3.2118 \times 10^{-24} \ \frac{\text{g}}{\text{atom}}$$

23.

Sig digs in all products are determined by the factor with the lowest precision. Sig digs in all sums are determined according to the addition rule.

silicon

$27.9769 \text{ u} \cdot 0.92223 = 25.8011 \text{ u}$

$28.9765 \text{ u} \cdot 0.04685 = 1.3585 \text{ u}$

$29.9738 \text{ u} \cdot 0.03092 = 0.92679 \text{ u}$

$$
\begin{array}{r}
25.8011 \text{ u} \\
1.3585 \text{ u} \\
+ 0.92679 \text{ u} \\
\hline
28.086 \text{ u}
\end{array}
$$

calcium

$39.9626 \text{ u} \cdot 0.96941 = 38.740 \text{ u}$

$41.9856 \text{ u} \cdot 0.00647 = 2.72 \text{ u}$

$$
\begin{array}{r}
38.740 \text{ u} \\
+ \ 0.272 \text{ u} \\
\hline
39.012 \text{ u}
\end{array}
$$

---

iron

53.9396 u·0.05845 = 3.153 u

55.9349 u·0.91754 = 51.323 u

56.9354 u·0.02119 = 1.206 u

57.9333 u·0.00282 = 0.163 u

  3.153 u

 51.323 u

  1.206 u

+ 0.163 u

 55.822 u

uranium

235.0439 u·0.007204 = 1.693 u

238.0508 u·0.992742 = 236.323 u

  1.693 u

+ 236.323 u

 238.016 u

24. a. For all items in problem 24, sig digs are determined as in problem 23.

$1 \cdot 14.0067 \ \dfrac{g}{mol} = 14.0067 \ \dfrac{g}{mol}$

$3 \cdot 1.0079 \ \dfrac{g}{mol} = 3.0237 \ \dfrac{g}{mol}$

  14.0067 g/mol

+ 3.0237 g/mol

  17.0304 g/mol

24. b.

$1 \cdot 12.011 \ \dfrac{g}{mol} = 12.011 \ \dfrac{g}{mol}$

$2 \cdot 15.9994 \ \dfrac{g}{mol} = 31.9988 \ \dfrac{g}{mol}$

  12.011 g/mol

+ 31.9988 g/mol

  44.010 g/mol

24. c.

$2 \cdot 35.4527 \ \dfrac{g}{mol} = 70.9054 \ \dfrac{g}{mol}$

24. d.

$1 \cdot 63.546 \ \dfrac{g}{mol} = 63.546 \ \dfrac{g}{mol}$

$1 \cdot 32.066 \ \dfrac{g}{mol} = 32.066 \ \dfrac{g}{mol}$

$4 \cdot 15.9994 \ \dfrac{g}{mol} = 63.9976 \ \dfrac{g}{mol}$

  63.546 g/mol

  32.066 g/mol

+ 63.9976 g/mol

  159.610 g/mol

24. e.

$1 \cdot 40.078 \ \dfrac{g}{mol} = 40.078 \ \dfrac{g}{mol}$

$2 \cdot 14.0067 \ \dfrac{g}{mol} = 28.0134 \ \dfrac{g}{mol}$

$4 \cdot 15.9994 \ \dfrac{g}{mol} = 63.9976 \ \dfrac{g}{mol}$

  40.078 g/mol

  28.0134 g/mol

+ 63.9976 g/mol

  132.089 g/mol

24. f.

$$12 \cdot 12.011 \, \frac{g}{mol} = 144.132 \, \frac{g}{mol}$$

$$22 \cdot 1.0079 \, \frac{g}{mol} = 22.174 \, \frac{g}{mol}$$

$$11 \cdot 15.9994 \, \frac{g}{mol} = 175.993 \, \frac{g}{mol}$$

144.132 g/mol
22.174 g/mol
+ 175.993 g/mol
342.30 g/mol

24. g.

$$2 \cdot 12.011 \, \frac{g}{mol} = 24.022 \, \frac{g}{mol}$$

$$6 \cdot 1.0079 \, \frac{g}{mol} = 6.0474 \, \frac{g}{mol}$$

$$1 \cdot 15.9994 \, \frac{g}{mol} = 15.9994 \, \frac{g}{mol}$$

24.022 g/mol
6.0474 g/mol
+ 15.9994 g/mol
46.069 g/mol

24. h.

$$3 \cdot 12.011 \, \frac{g}{mol} = 36.033 \, \frac{g}{mol}$$

$$8 \cdot 1.0079 \, \frac{g}{mol} = 8.0632 \, \frac{g}{mol}$$

36.033 g/mol
+ 8.0632 g/mol
44.096 g/mol

24. i.

$$1 \cdot 28.0855 \, \frac{g}{mol} = 28.0855 \, \frac{g}{mol}$$

$$2 \cdot 15.9994 \, \frac{g}{mol} = 31.9988 \, \frac{g}{mol}$$

28.0855 g/mol
+ 31.9988 g/mol
60.0843 g/mol

25. a. For all items in problem 25, sig digs are determined as in problem 23.

$1 \cdot 24.3050 \, u = 24.3050 \, u$

$2 \cdot 35.4527 \, u = 70.9054 \, u$

24.3050 u
+ 70.9054 u
95.2104 u

25. b.

$1 \cdot 40.078 \, u = 40.078 \, u$

$2 \cdot 14.0067 \, u = 28.0134 \, u$

$6 \cdot 15.9994 \, u = 95.9964 \, u$

40.078 u
28.0134 u
+ 95.9964 u
164.088 u

25. c.

$1 \cdot 32.066 \, u = 32.066 \, u$

$4 \cdot 15.9994 \, u = 63.9976 \, u$

32.066 u
+ 63.9976 u
96.064 u

25. d.

	63.546 u
$1 \cdot 63.546$ u $= 63.546$ u	32.066 u
$1 \cdot 32.066$ u $= 32.066$ u	+ 63.9976 u
$4 \cdot 15.9994$ u $= 63.9976$ u	159.610 u

25. e.

	10.811 u
$1 \cdot 10.811$ u $= 10.811$ u	+ 56.9952 u
$3 \cdot 18.9984$ u $= 56.9952$ u	67.806 u

25. f.

	12.011 u
$1 \cdot 12.011$ u $= 12.011$ u	+ 141.811 u
$4 \cdot 35.4527$ u $= 141.811$ u	153.822 u

26.

$$2.25 \text{ mol AgNO}_3 \cdot \frac{169.8731 \text{ g}}{\text{mol}} = 382 \text{ g}$$

(3 sig digs because the given value has 3)

27. a.

$$2.25 \text{ kg CCl}_4 \cdot \frac{1000 \text{ g}}{\text{kg}} \cdot \frac{\text{mol}}{153.822 \text{ g}} = 14.6 \text{ mol}$$

(3 sig digs because the given value has 3)

27. b.

$$14.63 \text{ mol} \cdot \frac{6.022 \times 10^{23} \text{ particles}}{\text{mol}} = 8.81 \times 10^{24} \text{ particles}$$

For CCl_4, particles = molecules, and each molecule contains 1 carbon atom, giving 8.81×10^{24} carbon atoms. (Computing moles in step 1 was an intermediate calculation for this one, thus an extra digit was retained to get 14.63.)

27. c.

From Table 1.5, 1.078% of carbon atoms are carbon-13. This gives $8.81 \times 10^{24} \cdot 0.01078 = 9.50 \times 10^{22}$ atoms of carbon-13.

28. a.

$$1.00 \text{ gal} \cdot \frac{3.785 \text{ L}}{1 \text{ gal}} \cdot \frac{1000 \text{ mL}}{1 \text{ L}} = 3785 \text{ mL}$$

$$\rho = \frac{m}{V} \rightarrow m = \rho V = 1.000 \frac{g}{mL} \cdot 3785 \text{ mL} = 3785 \text{ g}$$

$$3785 \text{ g} \cdot \frac{\text{mol}}{18.02 \text{ g}} = 2.10 \times 10^2 \text{ mol}$$

(Result written in scientific notation in order to show 3 sig digs.)

28. b.

$$210.0 \text{ mol} \cdot \frac{6.022 \times 10^{23} \text{ particles}}{\text{mol}} = 1.264 \times 10^{26} \text{ particles}$$

Particles = molecules, and there are 2 H atoms in each molecule. Doubling and rounding to 3 sig digs gives 2.53×10^{26} H atoms.

28. c.

From Table 1.5, H-2 is 0.0115% of all hydrogen. $2.53 \times 10^{26} \cdot 0.000115 = 2.91 \times 10^{22}$ atoms of H-2.

29.

Assuming a 100-gram sample:

$$38.7 \text{ g C} \cdot \frac{\text{mol}}{12.011 \text{ g}} = 3.222 \text{ mol C} \quad (3.222/3.222 = 1)$$

$$9.7 \text{ g H} \cdot \frac{\text{mol}}{1.0079 \text{ g}} = 9.62 \text{ mol H} \quad (9.62/3.222 = 2.99 \approx 3)$$

$$51.6 \text{ g O} \cdot \frac{\text{mol}}{15.9994 \text{ g}} = 3.225 \text{ mol O} \quad (3.225/3.222 \approx 1)$$

These ratios give an empirical formula of CH_3O.

$1 \cdot 12.011 \text{ u} + 3 \cdot 1.0079 \text{ u} + 1 \cdot 15.9994 \text{ u} = 31.0 \text{ u}$

$62.1 \text{ u} / 31.0 \text{ u} \approx 2$

Applying the factor of 2 to the empirical formula gives a molecular formula of $C_2H_6O_2$.

30. a.

$$\text{C: } \frac{43.910 \text{ g}}{47.593 \text{ g}} = 0.92261 = 92.261\% \text{ C}$$

$$\text{H: } \frac{3.683 \text{ g}}{47.593 \text{ g}} = 0.07739 = 7.739\% \text{ H}$$

30. b.

Assuming a 100-gram sample:

$$92.261 \text{ g C} \cdot \frac{\text{mol}}{12.011 \text{ g}} = 7.68 \text{ mol} \quad (7.68/7.68 = 1)$$

$$7.739 \text{ g H} \cdot \frac{\text{mol}}{1.0079 \text{ g}} = 7.68 \text{ mol} \quad (7.68/7.68 = 1)$$

These ratios give an empirical formula of CH.

30. c.

$1 \cdot 12.011 \text{ u} + 1 \cdot 1.0079 \text{ u} = 13.019 \text{ u}$

$78.11 \text{ u}/13.019 \text{ u} = 6$

Applying the factor of 6 to the empirical formula gives a molecular formula of C_6H_6.

31.

$$125.0 \text{ g HClO}_3 \cdot \frac{\text{mol}}{84.459 \text{ g}} \cdot \frac{6.022 \times 10^{23} \text{ particles}}{\text{mol}} = 8.913 \times 10^{23} \text{ particles}$$
$$\text{(molecules)}$$

32. a.

$1 \cdot 22.9898 \text{ u} + 1 \cdot 1.0079 \text{ u} + 1 \cdot 12.011 \text{ u} + 3 \cdot 15.9994 \text{ u} = 84.007 \text{ u}$

$22.9898/84.007 = 0.27367 \rightarrow 27.367\% \text{ Na}$

$1.0079/84.007 = 0.011998 \rightarrow 1.1998\% \text{ H}$

$12.011/84.007 = 0.14298 \rightarrow 14.298\% \text{ C}$

$(3 \cdot 15.9994)/84.007 = 0.57136 \rightarrow 57.136\% \text{ O}$

The sig digs in the formula mass are limited by the decimals in the mass of carbon. When this value came out with 5 sig digs, all percentages after that had 5 sig digs.

32. b.

$2 \cdot 22.9898 \text{ u} + 1 \cdot 15.9994 \text{ u} = 61.9790 \text{ u}$

$(2 \cdot 22.9898)/61.9790 = 0.741858 \rightarrow 74.1858\% \text{ Na}$

$(1 \cdot 15.9994)/61.9790 = 0.258142 \rightarrow 25.8142\% \text{ O}$

32. c.

$2 \cdot 55.847 \text{ u} + 3 \cdot 15.9994 \text{ u} = 159.692 \text{ u}$

$(2 \cdot 55.847)/159.692 = 0.69943 \rightarrow 69.943\% \text{ Fe}$

$(3 \cdot 15.9994)/159.692 = 0.300567 \rightarrow 30.0567\% \text{ O}$

The sig digs in the formula mass are limited by the decimals in the mass of iron. The atomic mass of iron also limits the iron percentage to 5 sig digs.

32. d.

$1 \cdot 107.8682$ u$+1 \cdot 14.0067$ u$+3 \cdot 15.9994$ u$=169.8731$ u

$(1 \cdot 107.8682)/169.8731 = 0.6349928 \rightarrow 63.49928\%$ Ag

$(1 \cdot 14.0067)/169.8731 = 0.0824539 \rightarrow 8.24539\%$ N

$(3 \cdot 15.9994)/169.8731 = 0.282553 \rightarrow 28.2553\%$ O

---

32. e.

$1 \cdot 40.078$ u$+4 \cdot 12.011$ u$+6 \cdot 1.0079$ u$+4 \cdot 15.9994$ u$=158.167$ u

$(1 \cdot 40.078)/158.167 = 0.25339 \rightarrow 25.339\%$ Ca

$(4 \cdot 12.011)/158.167 = 0.30375 \rightarrow 30.375\%$ C

$(6 \cdot 1.0079)/158.167 = 0.038234 \rightarrow 3.8234\%$ H

$(4 \cdot 15.9994)/158.167 = 0.404620 \rightarrow 40.4620\%$ O

The sig digs in the formula mass are limited to the third decimal by Ca and C. In the results, the sig digs for Ca, C, and H are limited by the atomic masses to 5. O has 6 sig digs in the mass, thus 6 in the result.

---

32. f.

$9 \cdot 12.011$ u$+8 \cdot 1.0079$ u$+4 \cdot 15.9994$ u$=180.160$ u

$(9 \cdot 12.011)/180.160 = 0.60002 \rightarrow 60.002\%$ C

$(8 \cdot 1.0079)/180.160 = 0.044756 \rightarrow 4.4756\%$ H

$(4 \cdot 15.9994)/180.160 = 0.355226 \rightarrow 35.5226\%$ O

33.

$1 \cdot 65.39$ u$+1 \cdot 32.066$ u$+4 \cdot 15.9994$ u$+14 \cdot 1.0079$ u$+7 \cdot 15.9994$ u$=287.56$ u

$14 \cdot 1.0079$ u$+7 \cdot 15.9994$ u$=126.1064$ u

$126.1064 / 287.56 = 0.43854$ (43.854%)

34.

Assuming a 100-gram sample:

22.65 g S$\cdot \dfrac{\text{mol}}{32.066 \text{ g}} = 0.7064$ mol $(0.7064/0.7064 = 1)$

32.38 g Na$\cdot \dfrac{\text{mol}}{22.9898 \text{ g}} = 1.408$ mol $(1.408/0.7064 \approx 2)$

44.99 g O$\cdot \dfrac{\text{mol}}{15.9994 \text{ g}} = 2.812$ mol $(2.812/0.7064 \approx 4)$

These ratios give an empirical formula of SNa_2O_4. As we see later, this is actually Na_2SO_4.

35.

$1 \cdot 12.011 \text{ u} + 2 \cdot 1.0079 \text{ u} + 1 \cdot 15.9994 \text{ u} = 30.026 \text{ u}$

$120.12 / 30.026 \approx 4$

Applying the factor of 4 to the empirical formula gives a molecular formula of $C_4H_8O_4$.

36. a.

Assuming a 100-gram sample:

$49.5 \text{ g C} \cdot \dfrac{\text{mol}}{12.011} = 4.12 \text{ mol} \quad (4.12/1.03 = 4)$

$5.15 \text{ g H} \cdot \dfrac{\text{mol}}{1.0079} = 5.11 \text{ mol} \quad (5.11/1.03 \approx 5)$

$28.9 \text{ g N} \cdot \dfrac{\text{mol}}{14.0067} = 2.06 \text{ mol} \quad (2.06/1.03 = 2)$

$16.5 \text{ g O} \cdot \dfrac{\text{mol}}{15.9994} = 1.03 \text{ mol} \quad (1.03/1.03 = 1)$

These ratios give an empirical formula of $C_4H_5N_2O$. We get the multiplier factor from the empirical formula mass:

$4 \cdot 12.011 \text{ u} + 5 \cdot 1.0079 \text{ u} + 2 \cdot 14.0067 \text{ u} + 1 \cdot 15.9994 \text{ u} = 97.096 \text{ u}$

$195 / 97.096 \approx 2$

This factor gives a molecular formula of $C_8H_{10}N_4O_2$.

36. b.

Assuming a 100-gram sample:

$75.69 \text{ g C} \cdot \dfrac{\text{mol}}{12.011} = 6.302 \text{ mol} \quad (6.302/0.9694 \approx 6.5) \rightarrow (6.5 \cdot 2 = 13)$

$8.80 \text{ g H} \cdot \dfrac{\text{mol}}{1.0079} = 8.73 \text{ mol} \quad (8.73/0.9694 \approx 9) \rightarrow (9 \cdot 2 = 18)$

$15.51 \text{ g O} \cdot \dfrac{\text{mol}}{15.9994} = 0.9694 \text{ mol} \quad (0.9694/0.9694 = 1) \rightarrow (1 \cdot 2 = 2)$

The values are all doubled to obtain whole number ratios. These ratios give an empirical formula of $C_{13}H_{18}O_2$. We get the multiplier factor from the empirical formula mass:

$13 \cdot 12.011 \text{ u} + 18 \cdot 1.0079 \text{ u} + 2 \cdot 15.9994 \text{ u} = 206.284 \text{ u}$

$206 / 206.284 \approx 1$

The factor of 1 indicates that the empirical and molecular formulas are the same.

36. c.

Assuming a 100-gram sample:

$$81.71 \text{ g C} \cdot \frac{\text{mol}}{12.011} = 6.803 \text{ mol} \quad (6.803/6.803 = 1) \rightarrow (1 \cdot 3 = 3)$$

$$18.29 \text{ g H} \cdot \frac{\text{mol}}{1.0079} = 18.15 \text{ mol} \quad (18.15/6.803 \approx 2.668) \rightarrow (2.668 \cdot 3 \approx 8)$$

The values are all tripled to obtain whole number ratios. These ratios give an empirical formula of C_3H_8. We get the multiplier factor from the empirical formula mass:

$3 \cdot 12.011 \text{ u} + 8 \cdot 1.0079 \text{ u} = 44.096 \text{ u}$

$44.096 / 44.096 = 1$

The factor of 1 indicates that the empirical and molecular formulas are the same.

---

36. d.

Assuming a 100-gram sample:

$$57.14 \text{ g C} \cdot \frac{\text{mol}}{12.011} = 4.757 \text{ mol} \quad (4.757/0.680 \approx 7) \rightarrow (7 \cdot 2 = 14)$$

$$6.16 \text{ g H} \cdot \frac{\text{mol}}{1.0079} = 6.11 \text{ mol} \quad (6.11/0.680 \approx 9) \rightarrow (9 \cdot 2 = 18)$$

$$9.52 \text{ g N} \cdot \frac{\text{mol}}{14.0067} = 0.680 \text{ mol} \quad (0.680/0.680 = 1) \rightarrow (1 \cdot 2 = 2)$$

$$27.18 \text{ g O} \cdot \frac{\text{mol}}{15.9994} = 1.699 \text{ mol} \quad (1.699/0.680 = 2.5) \rightarrow (2.5 \cdot 2 = 5)$$

The values are all doubled to obtain whole number ratios. These ratios give an empirical formula of $C_{14}H_{18}N_2O_5$. We get the multiplier factor from the empirical formula mass:

$14 \cdot 12.011 \text{ u} + 18 \cdot 1.0079 \text{ u} + 2 \cdot 14.0067 \text{ u} + 5 \cdot 15.9994 \text{ u} = 294.31 \text{ u}$

$294.302 / 294.31 \approx 1$

The factor of 1 indicates that the empirical and molecular formulas are the same.

---

36. e.

Assuming a 100-gram sample:

$$92.26 \text{ g C} \cdot \frac{\text{mol}}{12.011} = 7.681 \text{ mol} \quad (7.681/7.68 \approx 1)$$

$$7.74 \text{ g H} \cdot \frac{\text{mol}}{1.0079} = 7.68 \text{ mol} \quad (7.68/7.68 = 1)$$

These ratios give an empirical formula of CH. We get the multiplier factor from the empirical formula mass:

$1 \cdot 12.011 \text{ u} + 1 \cdot 1.0079 \text{ u} = 13.019 \text{ u}$

$26.038 / 13.019 = 2$

This factor gives a molecular formula of C_2H_2.

37.

$9.581/10.5 = 0.912 \rightarrow 91.2\%$ C

$0.919/10.5 = 0.0875 \rightarrow 8.75\%$ H

Assuming a 100-gram sample:

$$91.2 \text{ g C} \cdot \frac{\text{mol}}{12.011 \text{ g}} = 7.59 \text{ mol} \quad (7.59/7.59 = 1) \rightarrow (1 \cdot 7 = 7)$$

$$8.75 \text{ g H} \cdot \frac{\text{mol}}{1.0079 \text{ g}} = 8.68 \text{ mol} \quad (8.68/7.59 \approx 1.14) \rightarrow (1.14 \cdot 7 = 8)$$

The values are multiplied by 7 to obtain whole number ratios. These ratios give an empirical formula of C_7H_8. We get the multiplier factor from the empirical formula mass:

$7 \cdot 12.011 \text{ u} + 8 \cdot 1.0079 \text{ u} = 92.140 \text{ u}$

$92.140/92.140 = 1$

The factor of 1 indicates that the empirical and molecular formulas are the same.

Chapter 2

6. a.

$$\frac{1.098 \text{ Å}}{2} + \frac{0.741 \text{ Å}}{2} = 0.920 \text{ Å} \qquad \frac{|1.012 \text{ Å} - 0.920 \text{ Å}|}{1.012 \text{ Å}} \times 100\% = 9.09\%$$

6. b.

$$\frac{1.242 \text{ Å}}{2} + \frac{1.208 \text{ Å}}{2} = 1.225 \text{ Å} \qquad \frac{|1.128 \text{ Å} - 1.225 \text{ Å}|}{1.128 \text{ Å}} \times 100\% = 8.60\%$$

6. c.

$$\frac{1.242 \text{ Å}}{2} + \frac{1.208 \text{ Å}}{2} = 1.225 \text{ Å} \qquad \frac{|1.160 \text{ Å} - 1.225 \text{ Å}|}{1.160 \text{ Å}} \times 100\% = 5.60\%$$

6. d.

$$\frac{1.893 \text{ Å}}{2} + \frac{1.413 \text{ Å}}{2} = 1.653 \text{ Å} \qquad \frac{|1.570 \text{ Å} - 1.653 \text{ Å}|}{1.570 \text{ Å}} \times 100\% = 5.29\%$$

33.

$$\lambda = 94 \text{ nm} \cdot \frac{1 \text{ m}}{1 \times 10^9 \text{ nm}} = 9.4 \times 10^{-8} \text{ m}$$

$$E = \frac{hv}{\lambda} = \frac{6.626 \times 10^{-34} \text{ J} \cdot \text{s} \cdot 2.9979 \times 10^8 \frac{\text{m}}{\text{s}}}{9.4 \times 10^{-8} \text{ m}} = 2.11 \times 10^{-18} \text{ J} \cdot \frac{1 \text{ eV}}{1.60 \times 10^{-19} \text{ J}} = 13 \text{ eV}$$

34.

$$E = 5.09 \times 10^{-19} \text{ J}$$

$$E = \frac{hv}{\lambda} \rightarrow \lambda = \frac{hv}{E} = \frac{6.626 \times 10^{-34} \text{ J} \cdot \text{s} \cdot 2.9979 \times 10^8 \frac{\text{m}}{\text{s}}}{5.09 \times 10^{-19} \text{ J}} = 3.903 \times 10^{-7} \text{ m} \cdot \frac{1 \times 10^9 \text{ nm}}{1 \text{ m}} = 390 \text{ nm}$$

The result is stated with only 2 sig digs instead of three so we can see it in nanometers and compare it to the visible band of 700–400 nm. Doing so, the light is in the ultraviolet band and would thus not be visible.

37.

$$112 \text{ g CO}_2 \cdot \frac{\text{mol}}{44.01 \text{ g}} \cdot \frac{6.022 \times 10^{23} \text{ particles}}{\text{mol}} = 1.53 \times 10^{24} \text{ particles}$$

For CO_2, the particles are molecules and there is one atom of C in each molecule. Thus, this value is the number of carbon atoms.

38.

$2 \cdot 1.0079 \text{ u} + 1 \cdot 32.066 \text{ u} + 4 \cdot 15.9994 \text{ u} = 98.079 \text{ u}$

$2 \cdot 1.0079 \text{ u} / 98.079 \text{ u} = 0.020553 \rightarrow 2.0553\% \text{ H}$

$1 \cdot 32.066 \text{ u} / 98.079 \text{ u} = 0.32694 \rightarrow 32.694\% \text{ S}$

$4 \cdot 15.9994 \text{ u} / 98.079 \text{ u} = 0.65251 \rightarrow 65.251\% \text{ O}$

$35.0 \text{ g H}_2\text{SO}_4 \cdot \dfrac{\text{mol}}{98.079 \text{ g}} \cdot \dfrac{6.022 \times 10^{23} \text{ particles}}{\text{mol}} = 2.15 \times 10^{23} \text{ particles}$

Each particle is a molecule, and each molecule contains 2 H atoms, 1 S atom, and 4 O atoms. Thus, the numbers of atoms are 4.30×10^{23} H atoms, 2.15×10^{23} S atoms, and 8.60×10^{23} O atoms.

40.

The difference in mass between Br-79 and Br-81 is 2 u. The difference between the average mass 79.904 and 79 is 0.904. 0.904/2 = 0.452. If the average were 80, the proportion of each isotope would be 0.5. Instead the proportions are 0.452 and 0.548. Since the average mass of 79.904 is closer to 79 than it is to 81, the proportion of Br-79 is greater, and is 0.548. Thus, the closest of the choices given is 52%.

41.

$3.00 \text{ mol CaBr}_2 \cdot \dfrac{199.89 \text{ g}}{\text{mol}} = 599.7 \text{ g}$

$40.078 \text{ u} / 199.89 \text{ u} = 0.2005$ (proportion of calcium)

$0.2005 \cdot 599.7 \text{ g} = 1.20 \times 10^2 \text{ g}$

The result is stated in scientific notation in order to show 3 sig digs.

42.

Assuming a 100-gram sample:

$53.64 \text{ g Cl} \cdot \dfrac{\text{mol}}{35.4527 \text{ g}} = 1.513 \text{ mol} \quad (1.513 / 0.2522 = 6)$

$46.36 \text{ g W} \cdot \dfrac{\text{mol}}{183.85 \text{ g}} = 0.2522 \text{ mol} \quad (0.2522 / 0.2522 = 1)$

The 6:1 ratio gives an empirical formula of WCl_6.

Chapter 3

20. a.

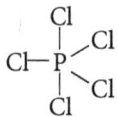

The Lewis structure is: $S{=}C{=}S$ Each bond is a double bond, so bond number is 2.

20. b.

The Lewis structure is:

$$
\begin{array}{c}
\overset{\displaystyle Cl}{\underset{\displaystyle Cl}{Cl{-}P}}\overset{\displaystyle {-}Cl}{\underset{\displaystyle {\diagdown}Cl}{}}
\end{array}
$$

Each bond is a single bond, so bond number is 1.

20. c.

The Lewis structure is: $\left[O{\cdots}\overset{..}{N}{\cdots}O \right]^{-}$ Each bond is composed of one part single bond and one part that is its share of the resonance structure bond. There is one resonance structure bond to be shared by two bonds, so the share of each is 0.5. Adding this to the single bond we get a bond number of 1.5.

22. a.

Be—F: 2.41, O—F: 0.54, C—F: 1.43

22. b.

F—F: 0, B—F: 1.94, S—O: 0.86

22. c.

O—Cl: 0.28, S—Cl: 0.58, C—P: 0.36

24.

$$
100.00 \text{ g CaCO}_3 \cdot \frac{\text{mol}}{100.087 \text{ g}} \cdot \frac{6.02214 \times 10^{23} \text{ particles}}{\text{mol}} = 6.0169 \times 10^{23} \text{ particles}
$$

In this ionic compound, the particles are formula units. There are 3 O atoms in each formula unit, so the number of O atoms is 1.8051×10^{24}.

25.

$1 \cdot 32.066 \text{ u} + 6 \cdot 18.9984 \text{ u} = 146.056 \text{ u}$

$1 \cdot 32.066 \text{ u} / 146.056 \text{ u} = 0.21955 \rightarrow 21.955\% \text{ S}$

$6 \cdot 18.9984 \text{ u} / 146.056 \text{ u} = 0.780457 \rightarrow 78.0457\% \text{ F}$

28.

$12.00 \text{ u} \cdot 0.9893 + 13.0034 \text{ u} \cdot 0.01078 = 12.01 \text{ u}$

$$
\text{percent difference} = \frac{\left| 12.01 \text{ u} - 12.01 \text{ u} \right|}{12.01 \text{ u}} \times 100\% = 0.00\%
$$

29.

$$\rho = \frac{m}{V} \rightarrow m = \rho V = 2.33 \frac{g}{cm^3} \cdot 1.000 \ cm^3 = 2.33 \ g$$

$$2.33 \ g \ MgCl_2 \cdot \frac{mol}{95.2104 \ g} \cdot \frac{6.022 \times 10^{23} \ particles}{mol} = 1.47 \times 10^{22} \ particles$$

In this ionic compound, the particles are formula units. There are 2 Cl atoms in each formula unit, so the number of Cl atoms is 2.94×10^{22}.

32.

Assuming a 100-gram sample:

$$12.84 \ g \ S \cdot \frac{mol}{32.066 \ g} = 0.4044 \ mol \ \ (0.4044/0.4011 \approx 1)$$

$$25.45 \ g \ Cu \cdot \frac{mol}{63.546 \ g} = 0.4011 \ mol \ \ (0.4011/0.4011 = 1)$$

$$25.63 \ g \ O \cdot \frac{mol}{15.9994 \ g} = 1.602 \ mol \ \ (1.602/0.4011 \approx 4)$$

$$36.07 \ g \ H_2O \cdot \frac{mol}{18.0152 \ g} = 2.002 \ mol \ \ (2.002/0.4011 \approx 5)$$

These ratios give an empirical formula of $CuSO_4 \cdot 5H_2O$. This is copper sulfate pentahydrate.

33.

$$E = 3.73 \times 10^{-19} \ J$$

$$E = \frac{hv}{\lambda} \rightarrow \lambda = \frac{hv}{E} = \frac{6.626 \times 10^{-34} \ J \cdot s \cdot 2.9979 \times 10^8 \ \frac{m}{s}}{3.73 \times 10^{-19} \ J} = 5.33 \times 10^{-7} \ m$$

$$5.33 \times 10^{-7} \ m \cdot \frac{10^9 \ nm}{1 \ m} = 533 \ nm$$

Chapter 4

23.

Formula: $C_3NO_2H_7$

$3 \cdot 12.011 \text{ u} / 89.094 \text{ u} = 0.40444 \rightarrow 40.444\% \text{ C}$

$1 \cdot 14.0067 \text{ u} / 89.094 \text{ u} = 0.15721 \rightarrow 15.721\% \text{ N}$

$2 \cdot 15.9994 \text{ u} / 89.094 \text{ u} = 0.35916 \rightarrow 35.916\% \text{ O}$

$7 \cdot 1.0079 \text{ u} / 89.094 \text{ u} = 0.079189 \rightarrow 7.9189\% \text{ H}$

24.

$$\lambda = 601 \text{ nm} \cdot \frac{1 \text{ m}}{1 \times 10^9 \text{ nm}} = 6.01 \times 10^{-7} \text{ m}$$

$$E = \frac{hv}{\lambda} = \frac{6.626 \times 10^{-34} \text{ J} \cdot \text{s} \cdot 2.9979 \times 10^8 \ \frac{\text{m}}{\text{s}}}{6.01 \times 10^{-7} \text{ m}} = 3.305 \times 10^{-19} \text{ J} \cdot \frac{1 \text{ eV}}{1.602 \times 10^{-19} \text{ J}} = 2.06 \text{ eV}$$

33. a.

$$65.6 \text{ g H}_2\text{O} \cdot \frac{\text{mol}}{18.02 \text{ g}} = 3.64 \text{ mol}$$

33. b.

$$1250 \text{ mg C}_6\text{H}_6\text{O}_6 \cdot \frac{\text{g}}{1000 \text{ mg}} \cdot \frac{\text{mol}}{174.110 \text{ g}} = 0.00718 \text{ mol}$$

33. c.

$$400 \text{ mg C}_9\text{H}_8\text{O}_4 \cdot \frac{\text{g}}{1000 \text{ mg}} \cdot \frac{\text{mol}}{180.160 \text{ g}} = 0.002 \text{ mol}$$

This result has been rounded to 1 sig dig since the given quantity (400 mg) has only 1 sig dig.

33. d.

$$14.0 \text{ kg KNO}_3 \cdot \frac{1000 \text{ g}}{1 \text{ kg}} \cdot \frac{\text{mol}}{101.103 \text{ g}} = 138 \text{ mol}$$

This result has been rounded to 3 sig digs since the given quantity (14.0 kg) has 3 sig digs.

33. e.

$$1050 \text{ g HClO}_4 \cdot \frac{\text{mol}}{100.4582 \text{ g}} = 10.5 \text{ mol}$$

This result has been rounded to 3 sig digs since the given quantity (1050 g) has 3 sig digs.

33. f.

$$953.00 \text{ g HF} \cdot \frac{\text{mol}}{20.0063 \text{ g}} = 47.635 \text{ mol}$$

This result has been rounded to 5 sig digs since the given quantity (953.00 g) has 5 sig digs.

34. a.

$$55 \text{ g HgS} \cdot \frac{\text{mol}}{232.66 \text{ g}} \cdot \frac{6.022 \times 10^{23} \text{ particles}}{\text{mol}} = 1.4 \times 10^{23} \text{ particles}$$

In this ionic compound, a particle is a formula unit. Each formula unit contains one Hg atom, so the number of Hg atoms is equal to the number of particles.

34. b.

$$3.00 \text{ kg Fe}_2\text{O}_3 \cdot \frac{1000 \text{ g}}{1 \text{ kg}} \cdot \frac{\text{mol}}{159.692 \text{ g}} \cdot \frac{6.022 \times 10^{23} \text{ particles}}{\text{mol}} = 1.13 \times 10^{25} \text{ particles}$$

In this ionic compound, a particle is a formula unit. Each formula unit contains two Fe atoms, so the number of Fe atoms is twice the number of particles, or 2.26×10^{25}.

34. c.

$$1.0000 \text{ mol CaCO}_3 \cdot \frac{6.0221 \times 10^{23} \text{ particles}}{\text{mol}} = 6.0221 \times 10^{23} \text{ particles}$$

In this ionic compound, a particle is a formula unit. Each formula unit contains one Ca atom, so the number of Ca atoms is equal to the number of particles. The Avogadro constant was written with 5 sig digs because the given quantity (1.0000 mol) contains 5 sig digs.

34. d

$$45 \text{ g Sr(NO}_2)_2 \cdot \frac{\text{mol}}{179.63 \text{ g}} \cdot \frac{6.022 \times 10^{23} \text{ particles}}{\text{mol}} = 1.5 \times 10^{23} \text{ particles}$$

In this ionic compound, a particle is a formula unit. Each formula unit contains one Sr atom, so the number of Sr atoms is equal to the number of particles.

34. e.

$$2.000 \text{ mol Na}_2\text{CrO}_4 \cdot \frac{6.022 \times 10^{23} \text{ particles}}{\text{mol}} = 1.2044 \times 10^{24} \text{ particles}$$

This result has one extra sig dig. In this ionic compound, a particle is a formula unit. Each formula unit contains two Na atoms, so the number of Na atoms is twice the number of particles, or 2.4088×10^{24}. Rounding to 4 sig digs gives 2.409×10^{24} Na atoms.

34. f.

$$5.05 \text{ kg Ca(CH}_3\text{COO)}_2 \cdot \frac{1000 \text{ g}}{1 \text{ kg}} \cdot \frac{\text{mol}}{158.167 \text{ g}} \cdot \frac{6.022 \times 10^{23} \text{ particles}}{\text{mol}} = 1.92 \times 10^{25} \text{ particles}$$

In this ionic compound, a particle is a formula unit. Each formula unit contains one Ca atom, so the number of Ca atoms is equal to the number of particles.

36.

53.9396 u·0.05845+55.9349 u·0.91754+56.9354 u·0.02119+57.9333 u·0.00282 =

$$3.1528 \text{ u} + 51.323 \text{ u} + 1.206 \text{ u} + 0.163 \text{ u} = 55.845 \text{ u}$$

The number of sig digs is driven by the addition rule, the result being limited to three decimal places.

40.

4.62 g + 0.776 g + 6.154 g = 11.55 g (sample mass)

4.62 g / 11.55 g = 0.400 → 40.0% C

0.776 g / 11.55 g = 0.0672 → 6.72% H

6.154 g / 11.55 g = 0.5328 → 53.28% O

The sig digs shown here result from the given quantities. (To add these percentages, decimals are limited by the addition rule to 1 decimal place, so the quantities would have to be rounded to 40.0%, 6.7%, and 53.3%. Arguably, this is also a legitimate way of expressing the result.)

Assuming a 100-gram sample:

$$40.0 \text{ g C} \cdot \frac{\text{mol}}{12.011 \text{ g}} = 3.33 \text{ mol} \quad (3.33/3.33 = 1)$$

$$6.72 \text{ g H} \cdot \frac{\text{mol}}{1.0079 \text{ g}} = 6.67 \text{ mol} \quad (6.67/3.33 = 2)$$

$$53.28 \text{ g O} \cdot \frac{\text{mol}}{15.9994 \text{ g}} = 3.330 \text{ mol} \quad (3.330/3.33 = 1)$$

These ratios give an empirical formula of CH_2O.

1·12.011 u + 2·1.0079 u + 1·15.9994 u = 30.026 u

180.157 u / 30.026 u ≈ 6

Applying this multiplier to the empirical formula gives a molecular formula of $C_6H_{12}O_2$.

Chapter 5

16.

$$AgNO_3 + NaBr \rightarrow AgBr + NaNO_3$$

$$3.5 \text{ mol } AgNO_3 \cdot \frac{1 \text{ mol } AgBr}{1 \text{ mol } AgNO_3} = 3.5 \text{ mol } AgBr$$

$$3.5 \text{ mol } AgBr \cdot \frac{187.772 \text{ g}}{\text{mol}} = 657 \text{ g}$$

Rounded to 2 sig digs, the answer is 660 g.

17.

$$2H_2 + O_2 \rightarrow 2H_2O$$

$$1575 \text{ mol } H_2 \cdot \frac{2 \text{ mol } H_2O}{2 \text{ mol } H_2} = 1575 \text{ mol } H_2O$$

$$1575 \text{ mol } H_2 \cdot \frac{1 \text{ mol } O_2}{2 \text{ mol } H_2} = 787.5 \text{ mol } O_2$$

18.

$$Al_2S_3 + 6H_2O \rightarrow 3H_2S + 2Al(OH)_3$$

$$2.290 \text{ kg } Al_2S_3 \cdot \frac{1000 \text{ g}}{1 \text{ kg}} \cdot \frac{\text{mol}}{150.161 \text{ g}} = 15.250 \text{ mol } Al_2S_3$$

$$15.250 \text{ mol } Al_2S_3 \cdot \frac{2 \text{ mol } Al(OH)_3}{1 \text{ mol } Al_2S_3} = 30.500 \text{ mol } Al(OH)_3$$

$$30.50 \text{ mol } Al(OH)_3 \cdot \frac{78.0034 \text{ g}}{\text{mol}} = 2379 \text{ g } Al(OH)_3$$

19.

$$Al(OH)_3 + 3HCl \rightarrow AlCl_3 + 3H_2O$$

19. a.

$$750 \text{ mg } Al(OH)_3 \cdot \frac{1 \text{ g}}{1000 \text{ mg}} \cdot \frac{\text{mol}}{78.0034 \text{ g}} = 0.00961 \text{ mol } Al(OH)_3$$

$$0.00961 \text{ mol } Al(OH)_3 \cdot \frac{3 \text{ mol } HCl}{1 \text{ mol } Al(OH)_3} = 0.0288 \text{ mol } HCl$$

Rounding this result to 2 sig digs gives 0.029 mol HCl.

19. b.

$$750 \text{ mg Al(OH)}_3 \cdot \frac{1 \text{ g}}{1000 \text{ mg}} \cdot \frac{\text{mol}}{78.0034 \text{ g}} = 0.00961 \text{ mol Al(OH)}_3$$

$$0.00961 \text{ mol Al(OH)}_3 \cdot \frac{3 \text{ mol H}_2\text{O}}{1 \text{ mol Al(OH)}_3} = 0.0288 \text{ mol H}_2\text{O}$$

$$0.0288 \text{ mol H}_2\text{O} \cdot \frac{18.02 \text{ g}}{\text{mol}} = 0.5198 \text{ g H}_2\text{O}$$

Rounding this result to 2 sig digs gives 0.52 g H_2O.

20.

$$\text{Fe}_2\text{O}_3 + 3\text{CO} \rightarrow 2\text{Fe} + 3\text{CO}_2$$

$$2.50 \times 10^4 \text{ kg Fe}_2\text{O}_3 \cdot \frac{1000 \text{ g}}{1 \text{ kg}} \cdot \frac{\text{mol}}{159.692 \text{ g}} = 156{,}600 \text{ mol Fe}_2\text{O}_3$$

$$156{,}600 \text{ mol Fe}_2\text{O}_3 \cdot \frac{2 \text{ mol Fe}}{1 \text{ mol Fe}_2\text{O}_3} = 313{,}100 \text{ mol Fe}$$

$$313{,}100 \text{ mol Fe} \cdot \frac{55.847 \text{ g}}{\text{mol}} = 1.75 \times 10^7 \text{ g Fe} \cdot \frac{1 \text{ kg}}{1000 \text{ g}} = 1.75 \times 10^4 \text{ kg Fe}$$

21.

$$\text{H}_2\text{SO}_4 + 2\text{NaOH} \rightarrow \text{Na}_2\text{SO}_4 + 2\text{H}_2\text{O}$$

21. a.

$$29.55 \text{ g NaOH} \cdot \frac{\text{mol}}{39.9971 \text{ g}} = 0.7388 \text{ mol NaOH}$$

$$44.11 \text{ g H}_2\text{SO}_4 \cdot \frac{\text{mol}}{98.079 \text{ g}} = 0.4497 \text{ mol H}_2\text{SO}_4$$

$$0.4497 \text{ mol H}_2\text{SO}_4 \cdot \frac{2 \text{ mol NaOH}}{1 \text{ mol H}_2\text{SO}_4} = 0.8994 \text{ mol NaOH required}$$

This much NaOH is not available, so NaOH is the limiting reactant.

---

21. b.

$$0.73880 \text{ mol NaOH} \cdot \frac{1 \text{ mol Na}_2\text{SO}_4}{2 \text{ mol NaOH}} = 0.36940 \text{ mol Na}_2\text{SO}_4$$

$$0.36940 \text{ mol Na}_2\text{SO}_4 \cdot \frac{142.043 \text{ g}}{\text{mol}} = 52.47 \text{ g Na}_2\text{SO}_4$$

---

21. c.

$$0.73880 \text{ mol NaOH} \cdot \frac{2 \text{ mol H}_2\text{O}}{2 \text{ mol NaOH}} = 0.73880 \text{ mol H}_2\text{O}$$

$$0.73880 \text{ mol H}_2\text{O} \cdot \frac{18.015 \text{ g}}{\text{mol}} = 13.31 \text{ g H}_2\text{O}$$

---

22. a.

$$350.0 \text{ mol } C_8H_{18} \cdot \frac{25 \text{ mol } O_2}{2 \text{ mol } C_8H_{18}} = 4375 \text{ mol } O_2$$

The result has been rounded to 4 sig digs as required by the given information.

---

22. b.

$$3.5 \text{ kg } O_2 \cdot \frac{1000 \text{ g}}{1 \text{ kg}} \cdot \frac{\text{mol}}{32.00 \text{ g}} = 109.4 \text{ mol } O_2$$

$$109.4 \text{ mol } O_2 \cdot \frac{18 \text{ mol } H_2O}{25 \text{ mol } O_2} = 78.77 \text{ mol } H_2O$$

$$78.77 \text{ mol } H_2O \cdot \frac{18.02 \text{ g}}{\text{mol}} \cdot \frac{1 \text{ kg}}{1000 \text{ g}} = 1.4 \text{ kg } H_2O$$

The result has been rounded to 2 sig digs as required by the given information.

---

22. c.

$$11.5 \text{ gal } C_8H_{18} \cdot \frac{3.785 \text{ L}}{1 \text{ gal}} \cdot \frac{1000 \text{ mL}}{1 \text{ L}} \cdot \frac{0.692 \text{ g}}{\text{mL}} = 30{,}120 \text{ g } C_8H_{18}$$

$$30{,}120 \text{ g } C_8H_{18} \cdot \frac{\text{mol}}{114.230 \text{ g}} = 263.7 \text{ mol } C_8H_{18}$$

$$263.7 \text{ mol } C_8H_{18} \cdot \frac{25 \text{ mol } O_2}{2 \text{ mol } C_8H_{18}} = 3296 \text{ mol } O_2$$

$$3296 \text{ mol } O_2 \cdot \frac{32.00 \text{ g}}{\text{mol}} \cdot \frac{1 \text{ kg}}{1000 \text{ g}} = 105 \text{ kg } O_2$$

The result has been rounded to 3 sig digs as required by the given information.

---

23. a.

$$855 \text{ g } NH_3 \cdot \frac{\text{mol}}{17.0304 \text{ g}} = 50.20 \text{ mol } NH_3$$

$$1750 \text{ g } O_2 \cdot \frac{\text{mol}}{32.00 \text{ g}} = 54.69 \text{ mol } O_2$$

$$50.20 \text{ mol } NH_3 \cdot \frac{5 \text{ mol } O_2}{4 \text{ mol } NH_3} = 62.75 \text{ mol } O_2 \text{ required.}$$

This much O_2 is not available, so O_2 is the limiting reactant.

---

23. b.

$$1750 \text{ g O}_2 \cdot \frac{\text{mol}}{32.00 \text{ g}} = 54.69 \text{ mol O}_2$$

$$54.69 \text{ mol O}_2 \cdot \frac{4 \text{ mol NO}}{5 \text{ mol O}_2} = 43.75 \text{ mol NO}$$

$$43.75 \text{ mol NO} \cdot \frac{30.006 \text{ g}}{\text{mol}} = 1313 \text{ g NO}$$

Rounding to 3 sig digs as required by the given information, we have 1310 g NO.

---

23. c.

$$\frac{1272 \text{ g}}{1313 \text{ g}} \cdot 100\% = 96.9\%$$

24.

$$2\text{Al(OH)}_3 + 3\text{H}_2\text{SO}_4 \rightarrow \text{Al}_2(\text{SO}_4)_3 + 6\text{H}_2\text{O}$$

24. a.

$$31.8 \text{ g H}_2\text{SO}_4 \cdot \frac{\text{mol}}{98.079 \text{ g}} = 0.3242 \text{ mol H}_2\text{SO}_4$$

$$25.4 \text{ g Al(OH)}_3 \cdot \frac{\text{mol}}{78.0034 \text{ g}} = 0.3256 \text{ mol Al(OH)}_3$$

$$0.3256 \text{ mol Al(OH)}_3 \cdot \frac{3 \text{ mol H}_2\text{SO}_4}{2 \text{ mol Al(OH)}_3} = 0.4884 \text{ mol H}_2\text{SO}_4 \text{ required.}$$

This much H_2SO_4 is not available, so H_2SO_4 is the limiting reactant.

---

24. b.

$$0.3242 \text{ mol H}_2\text{SO}_4 \cdot \frac{2 \text{ mol Al(OH)}_3}{3 \text{ mol H}_2\text{SO}_4} = 0.2161 \text{ mol Al(OH)}_3$$

$$0.2161 \text{ mol Al(OH)}_3 \cdot \frac{78.0034 \text{ g}}{\text{mol}} = 16.86 \text{ g Al(OH)}_3 \text{ are consumed}$$

$$25.4 \text{ g} - 16.9 \text{ g} = 8.5 \text{ g Al(OH)}_3 \text{ remaining}$$

---

24. c.

$$0.3242 \text{ mol H}_2\text{SO}_4 \cdot \frac{1 \text{ mol Al}_2(\text{SO}_4)_3}{3 \text{ mol H}_2\text{SO}_4} = 0.1081 \text{ mol Al}_2(\text{SO}_4)_3$$

$$0.1081 \text{ mol Al}_2(\text{SO}_4)_3 \cdot \frac{342.154 \text{ g}}{\text{mol}} = 37.0 \text{ g Al}_2(\text{SO}_4)_3 \text{ produced}$$

$$0.3242 \text{ mol H}_2\text{SO}_4 \cdot \frac{6 \text{ mol H}_2\text{O}}{3 \text{ mol H}_2\text{SO}_4} = 0.6484 \text{ mol H}_2\text{O}$$

$$0.6484 \text{ mol H}_2\text{O} \cdot \frac{18.02 \text{ g}}{\text{mol}} = 11.7 \text{ g H}_2\text{O produced}$$

---

25. a.

$$1.200 \times 10^3 \text{ kg N}_2\text{H}_4 \cdot \frac{1000 \text{ g}}{1 \text{ kg}} \cdot \frac{\text{mol}}{32.0368 \text{ g N}_2\text{H}_4} = 37{,}457 \text{ mol N}_2\text{H}_4$$

$$1.000 \times 10^3 \text{ kg } (\text{CH}_3)_2 \text{N}_2\text{H}_2 \cdot \frac{1000 \text{ g}}{1 \text{ kg}} \cdot \frac{\text{mol}}{60.099 \text{ g N}_2\text{H}_4} = 16{,}639 \text{ mol } (\text{CH}_3)_2 \text{N}_2\text{H}_2$$

$$4.500 \times 10^3 \text{ kg N}_2\text{O}_4 \cdot \frac{1000 \text{ g}}{1 \text{ kg}} \cdot \frac{\text{mol}}{92.0110 \text{ g N}_2\text{O}_4} = 48{,}907 \text{ mol N}_2\text{O}_4$$

$$37{,}457 \text{ mol N}_2\text{H}_4 \cdot \frac{1 \text{ mol } (\text{CH}_3)_2 \text{N}_2\text{H}_2}{2 \text{ mol N}_2\text{H}_4} = 18{,}729 \text{ mol } (\text{CH}_3)_2 \text{N}_2\text{H}_2$$

This much $(\text{CH}_3)_2 \text{N}_2\text{H}_2$ is not available, so $(\text{CH}_3)_2 \text{N}_2\text{H}_2$ is the limiting reactant between these two reactants. Next, we compare this with N_2O_4:

$$16{,}639 \text{ mol } (\text{CH}_3)_2 \text{N}_2\text{H}_2 \cdot \frac{3 \text{ mol N}_2\text{O}_4}{1 \text{ mol } (\text{CH}_3)_2 \text{N}_2\text{H}_2} = 49{,}917 \text{ mol N}_2\text{O}_4$$

This much N_2O_4 is not available, so N_2O_4 is the limiting reactant for the reaction, and was thus consumed first.

25. b.

$$48{,}907 \text{ mol N}_2\text{O}_4 \cdot \frac{8 \text{ mol H}_2\text{O}}{3 \text{ mol N}_2\text{O}_4} = 130{,}491 \text{ mol H}_2\text{O}$$

$$130{,}491 \text{ mol H}_2\text{O} \cdot \frac{18.015 \text{ g}}{\text{mol}} \cdot \frac{1 \text{ kg}}{1000 \text{ g}} = 2349 \text{ kg H}_2\text{O}$$

25. c.

$$48{,}907 \text{ mol N}_2\text{O}_4 \cdot \frac{6 \text{ mol N}_2}{3 \text{ mol N}_2\text{O}_4} = 97{,}814 \text{ mol N}_2$$

$$97{,}814 \text{ mol N}_2 \cdot \frac{6.0221 \times 10^{23} \text{ particles}}{\text{mol}} = 5.890 \times 10^{28} \text{ particles N}_2$$

Since there are 2 N atoms per molecule, this gives 1.178×10^{29} N atoms.

26. a.

$$542 \text{ g SiO}_2 \cdot \frac{\text{mol}}{60.0843 \text{ g}} = 9.021 \text{ mol}$$

$$9.021 \text{ mol SiO}_2 \cdot \frac{6 \text{ mol HF}}{1 \text{ mol SiO}_2} = 54.1 \text{ mol HF}$$

26. b.

$$4.25 \text{ mol HF} \cdot \frac{1 \text{ mol H}_2\text{SiF}_6}{6 \text{ mol HF}} = 0.7083 \text{ mol H}_2\text{SiF}_6$$

$$0.7083 \text{ mol H}_2\text{SiF}_6 \cdot \frac{144.0917 \text{ g}}{\text{mol}} = 102 \text{ g H}_2\text{SiF}_6$$

26. c.

$$2.0 \text{ gal} \cdot \frac{3.785 \text{ L}}{1 \text{ gal}} \cdot \frac{1000 \text{ mL}}{1 \text{ L}} \cdot \frac{0.998 \text{ g}}{\text{mL}} = 7555 \text{ g H}_2\text{O}$$

$$7555 \text{ g H}_2\text{O} \cdot \frac{\text{mol}}{18.015 \text{ g}} = 419 \text{ mol H}_2\text{O}$$

$$419 \text{ mol H}_2\text{O} \cdot \frac{6 \text{ mol HF}}{2 \text{ mol H}_2\text{O}} = 1300 \text{ mol HF}$$

The result has been rounded to 2 sig digs as required.

26. d.

$$2.50 \text{ kg HF} \cdot \frac{1000 \text{ g}}{1 \text{ kg}} \cdot \frac{\text{mol}}{20.0063 \text{ g}} = 125.0 \text{ mol HF}$$

$$1.205 \text{ kg SiO}_2 \cdot \frac{1000 \text{ g}}{1 \text{ kg}} \cdot \frac{\text{mol}}{60.0843 \text{ g}} = 20.06 \text{ mol SiO}_2$$

$$20.06 \text{ mol SiO}_2 \cdot \frac{6 \text{ mol HF}}{1 \text{ mol SiO}_2} = 120.3 \text{ mol HF}$$

More HF than this is available, so the limiting reactant is SiO_2.

26. e.

$$20.06 \text{ mol SiO}_2 \cdot \frac{1 \text{ mol H}_2\text{SiF}_6}{1 \text{ mol SiO}_2} = 20.06 \text{ mol H}_2\text{SiF}_6$$

$$20.06 \text{ mol H}_2\text{SiF}_6 \cdot \frac{144.0917 \text{ g}}{\text{mol}} = 2890 \text{ g} \cdot \frac{1 \text{ kg}}{1000 \text{ g}} = 2.89 \text{ kg H}_2\text{SiF}_6$$

26. f.

$$\frac{2.657 \text{ kg}}{2.890 \text{ kg}} \cdot 100\% = 91.9\%$$

27.

$$C_6H_6 + Br_2 \rightarrow C_6H_5Br + HBr$$

27. a.

$$45.0 \text{ g C}_6\text{H}_6 \cdot \frac{\text{mol}}{78.113 \text{ g}} = 0.5761 \text{ mol C}_6\text{H}_6$$

$$97.5 \text{ g Br}_2 \cdot \frac{\text{mol}}{159.808 \text{ g}} = 0.6101 \text{ mol Br}_2$$

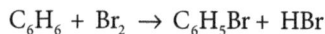

$$0.5761 \text{ mol C}_6\text{H}_6 \cdot \frac{1 \text{ mol Br}_2}{1 \text{ mol C}_6\text{H}_6} = 0.5761 \text{ mol Br}_2$$

More Br_2 than this is available, so C_6H_6 is the limiting reactant.

$$0.5761 \text{ mol } C_6H_6 \cdot \frac{1 \text{ mol } C_6H_5Br}{1 \text{ mol } C_6H_6} = 0.5761 \text{ mol } C_6H_5Br$$

$$0.5761 \text{ mol } C_6H_5Br \cdot \frac{157.010 \text{ g}}{\text{mol}} = 90.5 \text{ g } C_6H_5Br$$

---

27. b.

$$\frac{63.25 \text{ g}}{90.5 \text{ g}} \cdot 100\% = 69.9\%$$

---

28. a.

$$2.85 \times 10^{-6} \text{ mol } O_3 \cdot \frac{2 \text{ mol NaI}}{1 \text{ mol } O_3} = 5.70 \times 10^{-6} \text{ mol NaI}$$

---

28. b.

$$3.00 \text{ g } O_3 \cdot \frac{\text{mol}}{48.00 \text{ g}} = 0.06250 \text{ mol } O_3$$

$$0.06250 \text{ mol } O_3 \cdot \frac{2 \text{ mol NaI}}{1 \text{ mol } O_3} = 0.1250 \text{ mol NaI}$$

$$0.1250 \text{ mol NaI} \cdot \frac{149.08 \text{ g}}{\text{mol}} = 18.6 \text{ g NaI}$$

---

28. c.

$$1455 \text{ g NaI} \cdot \frac{\text{mol}}{149.8943 \text{ g}} = 9.7068 \text{ mol NaI}$$

$$250.0 \text{ g } O_3 \cdot \frac{\text{mol}}{47.998 \text{ g}} = 5.2085 \text{ mol } O_3$$

$$5.2086 \text{ mol } O_3 \cdot \frac{2 \text{ mol NaI}}{1 \text{ mol } O_3} = 10.417 \text{ mol NaI}$$

This much NaI is not available, so NaI is the limiting reactant.

$$9.7068 \text{ mol NaI} \cdot \frac{1 \text{ mol } I_2}{2 \text{ mol NaI}} = 4.8534 \text{ mol } I_2$$

$$4.8534 \text{ mol } I_2 \cdot \frac{253.809 \text{ g}}{\text{mol}} = 1232 \text{ g } I_2$$

This result is shown in scientific notation in order to show 4 sig digs.

---

28. d.

$$4.8534 \text{ mol } I_2 \cdot \frac{6.0221 \times 10^{23} \text{ particles}}{\text{mol}} = 2.923 \times 10^{24} \text{ particles}$$

Each particle is a molecules containing two I atoms, so there are 5.846×10^{24} I atoms.

---

29. a.

16.81 mg / 25.0 mg = 0.672 → 67.2% C

1.736 mg / 25.0 g = 0.0694 → 6.94% H

3.015 mg / 25.0 g = 0.121 → 12.1% N

3.445 mg / 25.0 g = 0.138 → 13.8% O

---

29. b.

Assuming a 100-gram sample:

$$67.2 \text{ g C} \cdot \frac{\text{mol}}{12.011 \text{ g}} = 5.59 \text{ mol} \quad (5.59/0.863 \approx 6.5) \rightarrow 6.5 \cdot 2 = 13$$

$$6.94 \text{ g H} \cdot \frac{\text{mol}}{1.0079 \text{ g}} = 6.89 \text{ mol} \quad (6.89/0.863 \approx 8) \rightarrow 8 \cdot 2 = 16$$

$$12.1 \text{ g N} \cdot \frac{\text{mol}}{14.0067 \text{ g}} = 0.864 \text{ mol} \quad (0.864/0.863 \approx 1) \rightarrow 1 \cdot 2 = 2$$

$$13.8 \text{ g O} \cdot \frac{\text{mol}}{15.9994 \text{ g}} = 0.863 \text{ mol} \quad (0.863/0.863 = 1) \rightarrow 1 \cdot 2 = 2$$

Proportions are doubled to give whole numbers. These ratios give an empirical formula of $C_{13}H_{16}N_2O_2$. The formula mass of this formula is 232.82 u, the same as the given molar mass. Thus, this is the molecular formula.

---

29. c.

$$3.0 \text{ mg } C_{13}H_{16}N_2O_2 \cdot \frac{1 \text{ g}}{1000 \text{ mg}} \cdot \frac{\text{mol}}{232 \text{ g}} \cdot \frac{6.022 \times 10^{23} \text{ particles}}{\text{mol}} = 7.79 \times 10^{18} \text{ particles}$$

Each particle (molecule) contains 13 C atoms, giving 1.0×10^{20} C atoms.

30.

$$E = 2.271 \text{ eV} \cdot \frac{1.60218 \times 10^{-19} \text{ J}}{1 \text{ eV}} = 3.6386 \times 10^{-19} \text{ J}$$

$$E = \frac{hv}{\lambda} \rightarrow \lambda = \frac{hv}{E} = \frac{6.626 \times 10^{-34} \text{ J} \cdot \text{s} \cdot 2.9979 \times 10^{8} \frac{\text{m}}{\text{s}}}{3.6386 \times 10^{-19} \text{ J}} = 5.459 \times 10^{-7} \text{ m} \cdot \frac{1 \times 10^{9} \text{ nm}}{1 \text{ m}} = 545.9 \text{ nm}$$

33.

27.9769 u · 0.92223 + 28.9765 u · 0.04685 + 29.9738 u · 0.03092 =

$$25.801 \text{ u} + 1.358 \text{ u} + 0.9268 \text{ u} = 28.085 \text{ u}$$

Chapter 6

5.

Since we are using the density in a calculation with the acceleration of gravity and pressure, we must use MKS units.

$$\rho = 0.998 \ \frac{g}{mL} \cdot \frac{1 \ kg}{1000 \ g} \cdot \frac{1000 \ mL}{1 \ L} \cdot \frac{1000 \ L}{m^3} = 998 \ \frac{kg}{m^3}$$

$$g = 9.80 \ \frac{m}{s^2}$$

$$P = 101,325 \ Pa$$

$$P = \rho g h \rightarrow h = \frac{P}{\rho g} = \frac{101,325 \ Pa}{998 \ \frac{kg}{m^3} \cdot 9.80 \ \frac{m}{s^2}} = 10.4 \ m$$

6.

$$1850 \ psi \cdot \frac{6894.8 \ Pa}{1 \ psi} = 12,755,000 \ Pa$$

Rounding to 3 sig digs, we have 12,800,000 Pa.

$$12,800,000 \ Pa \cdot \frac{1 \ kPa}{1000 \ Pa} = 12,800 \ kPa$$

$$12,755,000 \ Pa \cdot \frac{1 \ atm}{101,325 \ Pa} = 125.89 \ atm$$

Rounding to 3 sig digs, we have 126 atm.

$$125.89 \ atm \cdot \frac{760 \ Torr}{1 \ atm} = 95,700 \ Torr$$

$$12,800,000 \ Pa \cdot \frac{1 \ bar}{100,000 \ Pa} = 128 \ bar$$

7.

Using the symbol F_w to represent the skater's weight:

$$l = 2.9 \ mm \cdot \frac{1 \ m}{1000 \ mm} = 0.0029 \ m$$

$$w = 57 \ mm \cdot \frac{1 \ m}{1000 \ mm} = 0.057 \ m$$

$$F_w = 200.0 \ lb \cdot \frac{4.448 \ N}{1 \ lb} = 889.6 \ N$$

$$A = l \cdot w = 0.0029 \ m \cdot 0.0057 \ m = 0.0001653 \ m^2$$

$$P = \frac{F}{A} = \frac{889.6 \ N}{0.0001653 \ m^2} = 5,382,000 \ Pa$$

$$5,382,000 \ Pa \cdot \frac{1 \ atm}{101,325 \ Pa} = 53 \ atm$$

---

19. a.

$n = 3.47 \text{ mol}$

$\Delta T = T_f - T_i = 55°C - 25°C = 30°C = 30 \text{ K (with 2 sig digs)}$

$C = 0.0752 \; \dfrac{kJ}{mol \cdot K}$

$Q = Cn\Delta T = 0.0752 \; \dfrac{kJ}{mol \cdot K} \cdot 3.47 \text{ mol} \cdot 30 \text{ K} = 7.8 \text{ kJ}$

This value has been rounded to 2 sig digs as the given temperature data require.

19. b.

$Q = 3.65 \times 10^5 \text{ kJ}$

$H_v = 40.7 \; \dfrac{kJ}{mol}$

$Q = mH_v \rightarrow m = \dfrac{Q}{H_v} = \dfrac{3.65 \times 10^5 \text{ kJ}}{40.7 \; \dfrac{kJ}{mol}} = 8970 \text{ mol}$

$8970 \text{ mol} \cdot \dfrac{18.02 \text{ g}}{mol} \cdot \dfrac{1 \text{ kg}}{1000 \text{ g}} = 162 \text{ kg}$

19. c.

$m = 12.00 \text{ kg} \cdot \dfrac{1000 \text{ g}}{1 \text{ kg}} = 12{,}000 \text{ g (with 4 sig digs)}$

$H_f = 6.01 \; \dfrac{kJ}{mol}$

$n = 12{,}000 \text{ g} \cdot \dfrac{mol}{18.015 \text{ g}} = 666.11 \text{ mol}$

$Q = nH_f = 666.11 \text{ mol} \cdot 6.01 \; \dfrac{kJ}{mol} = 4003 \text{ kJ}$

To round this value to 3 sig digs, we must write it in scientific notation, giving 4.00×10^3 kJ.

19. d.

$Q = 7.32$ kJ

$\Delta T = T_f - T_i = 30.0°C - 20.0°C = 10.0°C = 10.0$ K

$C = 0.0752 \dfrac{\text{kJ}}{\text{mol} \cdot \text{K}}$

$\rho = 0.9970 \dfrac{\text{g}}{\text{cm}^3}$

$Q = Cn\Delta T \rightarrow n = \dfrac{Q}{C\Delta T} = \dfrac{7.32 \text{ kJ}}{0.0752 \dfrac{\text{kJ}}{\text{mol} \cdot \text{K}} \cdot 10.0 \text{ K}} = 9.734$ mol

$9.734 \text{ mol} \cdot \dfrac{18.015 \text{ g}}{\text{mol}} = 175.4$ g

$\rho = \dfrac{m}{V} \rightarrow V = \dfrac{m}{\rho} = \dfrac{175.4 \text{ g}}{0.9970 \dfrac{\text{g}}{\text{cm}^3}} = 176 \text{ cm}^3 = 176$ mL

---

19. e.

$m = 5.00$ kg

$\Delta T_{ice} = T_f - T_i = 0.0°C - (-25.0°C) = 25.0°C = 25.0$ K

$C_{ice} = 0.0364 \dfrac{\text{kJ}}{\text{mol} \cdot \text{K}}$

$H_f = 6.01 \dfrac{\text{kJ}}{\text{mol}}$

Begin by computing the number of moles:

$5.00 \text{ kg H}_2\text{O} \cdot \dfrac{1000 \text{ g}}{1 \text{ kg}} \cdot \dfrac{\text{mol}}{18.015 \text{ g}} = 277.5$ mol

Next compute the heat required to raise the temperature of the ice from $-25°C$ to $0°C$:

$Q = Cn\Delta T = 0.0364 \dfrac{\text{kJ}}{\text{mol} \cdot \text{K}} \cdot 277.5 \text{ mol} \cdot 25.0 \text{ K} = 252.5$ kJ

This value has one extra sig digs which we will round off later. Next compute the heat required to melt the ice:

$Q = nH_f = 277.5 \text{ mol} \cdot 6.10 \dfrac{\text{kJ}}{\text{mol}} = 1668$ kJ

This value also has one extra sig dig. The answer comes when we add these heat values (using the addition rule for the sig digs). To do this, we need to first round off each value to the correct number of sig digs, then add them using the rule.

1670 kJ

+ 253 kJ

1920 kJ

---

20. a.

The dotted line at atmospheric pressure (1 atm) indicates sublimation occurs at approximately −75°C.

20. b.

Using the sample scale as a reference, we estimate the triple point pressure to be 4 atm. Converting to Torr,

$$4 \text{ atm} \cdot \frac{760 \text{ Torr}}{1 \text{ atm}} = 3000 \text{ Torr}$$

This value is rounded to 1 sig dig. The triple-point temperature is approximately −55°C, which is 218 K.

20. c.

$$7600 \text{ Torr} \cdot \frac{1 \text{ atm}}{760 \text{ Torr}} = 10 \text{ atm}$$

At this pressure, the temperatures at the solid-liquid and liquid-gas boundaries are approximately −55°C and −35°C, respectively.

20. d.

This is the pressure at the tip of the liquid-gas curve, approximately 40 atm.

20. e.

Again using the logarithmic scale for reference, a vertical line at −20°C crosses the gas-liquid and liquid-solid curves at approximately 15 atm and 1000 atm, respectively.

20. f.

The vapor pressure is the pressure at the liquid-gas curve at a given temperature, or approximately 35 atm in this case.

20. g.

That pressure is above the solid-liquid curve, so the gas solidifies.

24. a.

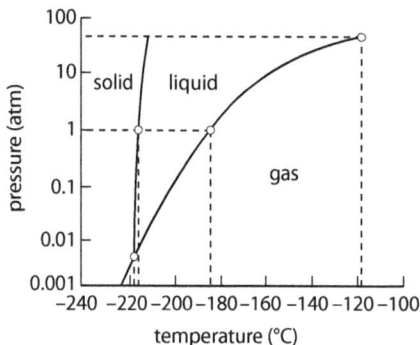

24. b.

The slope of the solid-liquid curve from the triple point to atmospheric pressure is positive, not negative as it is with water. This implies that solid oxygen is denser than liquid oxygen and will not float the way water ice does.

24. c.

At atmospheric pressure, heating moves the conditions of solid oxygen across the solid-liquid curve, and thus heating solid O_2 causes it to melt.

25.

The given pressure of 594 Torr falls between the pressures of 567.7 Torr and 611.6 Torr shown in Table B-4. Thus, the boiling point falls between the temperatures associated with these vapor pressures (92°C and 94°C), putting the boiling point of water at approximately 93°C.

27. a.

$$5.00 \text{ mol } C_8H_{18} \cdot \frac{25 \text{ mol } O_2}{2 \text{ mol } C_8H_{18}} = 62.5 \text{ mol } O_2$$

27. b.

$$\rho = 0.692 \ \frac{g}{mL}$$

$$1.00 \text{ gal } C_8H_{18} \cdot \frac{3.785 \text{ L}}{gal} \cdot \frac{1000 \text{ mL}}{L} = 3785 \text{ mL}$$

$$\rho = \frac{m}{V} \rightarrow m = \rho V = 0.692 \ \frac{g}{mL} \cdot 3785 \text{ mL} = 2619 \text{ g}$$

$$2619 \text{ g } C_8H_{18} \cdot \frac{mol}{114.23 \text{ g}} = 22.93 \text{ mol}$$

$$22.93 \text{ mol } C_8H_{18} \cdot \frac{25 \text{ mol } O_2}{2 \text{ mol } C_8H_{18}} = 286.6 \text{ mol } O_2$$

$$286.6 \text{ mol } O_2 \cdot \frac{32.00 \text{ g}}{mol} = 9170 \text{ g}$$

An extra sig dig was used during the calculation, and the result was rounded to 3 sig digs as the given information requires.

29. a.

$$2.50 \text{ g Zn} \cdot \frac{mol}{65.39 \text{ g}} = 0.03823 \text{ mol Zn}$$

$$3.00 \text{ g AgNO}_3 \cdot \frac{mol}{169.9 \text{ g}} = 0.01766 \text{ mol AgNO}_3$$

$$0.03823 \text{ mol Zn} \cdot \frac{2 \text{ mol AgNO}_3}{1 \text{ mol Zn}} = 0.07646 \text{ mol AgNO}_3$$

This much $AgNO_3$ is not available, so $AgNO_3$ is the limiting reactant.

29. b.

$$0.01766 \text{ mol AgNO}_3 \cdot \frac{1 \text{ mol Zn(NO}_3)_2}{2 \text{ mol AgNO}_3} = 0.00883 \text{ mol Zn(NO}_3)_2$$

$$0.00883 \text{ mol Zn(NO}_3)_2 \cdot \frac{189.40 \text{ g}}{\text{mol}} = 1.67 \text{ g}$$

---

29. c.

$$0.01766 \text{ mol AgNO}_3 \cdot \frac{1 \text{ mol Zn}}{2 \text{ mol AgNO}_3} = 0.00883 \text{ mol Zn}$$

$$0.03823 \text{ mol} - 0.00883 \text{ mol} = 0.0294 \text{ mol Zn remaining}$$

$$0.0294 \text{ mol Zn} \cdot \frac{65.39 \text{ g}}{\text{mol}} = 1.92 \text{ g}$$

30.

Assuming a 100-gram sample:

$$75.69 \text{ g C} \cdot \frac{\text{mol}}{12.011 \text{ g}} = 6.302 \text{ mol} \quad (6.302/0.969 \approx 6.5) \rightarrow 6.5 \cdot 2 = 13$$

$$8.80 \text{ g H} \cdot \frac{\text{mol}}{1.0079 \text{ g}} = 8.731 \text{ mol} \quad (8.731/0.969 \approx 9) \rightarrow 9 \cdot 2 = 18$$

$$15.51 \text{ g O} \cdot \frac{\text{mol}}{15.9994 \text{ g}} = 0.969 \text{ mol} \quad (0.969/0.969 = 1) \rightarrow 1 \cdot 2 = 2$$

Proportions are doubled to give whole numbers. These ratios give an empirical formula of $C_{13}H_{18}O_2$. The formula mass of this formula is 206.284 u, the same as the given molar mass. Thus, this is the molecular formula.

Chapter 7

1. a.

$V = 2.00$ L

$T = 22.0°C$

$P_1 = 0.974$ atm

$P_2 = 0.772$ atm

$$P_1V_1 = P_2V_2 \rightarrow V_2 = V_1 \cdot \frac{P_1}{P_2} = 2.00 \text{ L} \cdot \frac{0.974 \text{ atm}}{0.772 \text{ atm}} = 2.52 \text{ L}$$

---

1. b.

$V_1 = 2.00$ L

$V_2 = 3.14$ L

$T_1 = 22.0°C \rightarrow 22.0 + 273.2 = 295.2$ K

$P = 0.974$ atm

$$\frac{V_1}{T_1} = \frac{V_2}{T_2} \rightarrow T_2 = T_1 \cdot \frac{V_2}{V_1} = 295.2 \text{ K} \cdot \frac{3.14 \text{ L}}{2.00 \text{ L}} = 463 \text{ K}$$

2.

$V_1 = 3.5 \text{ m}^3$

$V_2 = 1.25 \text{ m}^3$

$P_1 = 0.275$ bar

$$P_1V_1 = P_2V_2 \rightarrow P_2 = P_1 \cdot \frac{V_1}{V_2} = 0.275 \text{ bar} \cdot \frac{3.5 \text{ m}^3}{1.25 \text{ m}^3} = 0.77 \text{ bar}$$

3.

Stage 1: Volume at the end of this stage is V_2.

$V_1 = 2225 \text{ cm}^3$

$T_1 = 37.00°C \rightarrow 37.00 + 273.15 = 310.15$ K

$T_2 = -15°C \rightarrow -15 + 273.2 = 258.2$ K

$$\frac{V_1}{T_1} = \frac{V_2}{T_2} \rightarrow V_2 = V_1 \cdot \frac{T_2}{T_1} = 2225 \text{ cm}^3 \cdot \frac{258.2 \text{ K}}{310.15 \text{ K}} = 1852 \text{ cm}^3$$

Stage 2: Volumes at beginning and end of this stage are V_2 and V_3.

$P_2 = 655$ Torr

$P_3 = 603$ Torr

$V_2 = 1852 \text{ cm}^3$

$$P_2V_2 = P_3V_3 \rightarrow V_3 = V_2 \cdot \frac{P_2}{P_3} = 1852 \text{ cm}^3 \cdot \frac{655 \text{ Torr}}{603 \text{ Torr}} = 2010 \text{ cm}^3$$

9.

$n = 100.0$ mol

$T = 0°C = 273.15$ K

$P = 100$ kPa

$R = 8.314 \, \dfrac{L \cdot kPa}{mol \cdot K}$

$$PV = nRT \rightarrow V = \frac{nRT}{P} = \frac{100.0 \text{ mol} \cdot 8.314 \, \dfrac{L \cdot kPa}{mol \cdot K} \cdot 273.15 \text{ K}}{100 \text{ kPa}} = 2271 \text{ L}$$

10.

$n = 0.105$ mol

$V = 1750 \text{ mL} \cdot \dfrac{1 \text{ L}}{1000 \text{ mL}} = 1.75 \text{ L}$

$P = 15.0 \text{ bar} \cdot \dfrac{100 \text{ kPa}}{1 \text{ bar}} = 1500 \text{ kPa}$

$R = 8.314 \, \dfrac{L \cdot kPa}{mol \cdot K}$

$$PV = nRT \rightarrow T = \frac{PV}{nR} = \frac{1500 \text{ kPa} \cdot 1.75 \text{ L}}{0.105 \text{ mol} \cdot 8.314 \, \dfrac{L \cdot kPa}{mol \cdot K}} = 3007 \text{ K}$$

With 3 sig digs this is 3010 K, with the most precise digit in the 10s column.

After the additions below, we will need to round to the 10s column.

$T_C = T_K - 273 = 3007 - 273 = 2734°C$

$T_F = \dfrac{9}{5} T_C + 32 = \dfrac{9}{5} \cdot 2734 + 32 = 4953°F$

Rounding to the 10s column, we have 4950°F.

11.

$m = 2.10$ g

$P = 145$ Torr

$T = 276$ K

$R = 62.36 \, \dfrac{L \cdot Torr}{mol \cdot K}$

$2.10 \text{ g CH}_4 \cdot \dfrac{mol}{16.043 \text{ g}} = 0.1309 \text{ mol} = n$

$$PV = nRT \rightarrow V = \frac{nRT}{P} = \frac{0.1309 \text{ mol} \cdot 62.36 \, \dfrac{L \cdot Torr}{mol \cdot K} \cdot 276 \text{ K}}{145 \text{ Torr}} = 15.5 \text{ L}$$

12.

$V = 1.75$ L

$P = 0.922 \text{ bar} \cdot \dfrac{100 \text{ kPa}}{1 \text{ bar}} = 922 \text{ kPa}$

$T = 5°C \rightarrow 5°C + 273.15 = 278.2$ K

$R = 8.314 \; \dfrac{L \cdot kPa}{mol \cdot K}$

$PV = nRT \rightarrow n = \dfrac{PV}{RT} = \dfrac{922 \text{ kPa} \cdot 1.75 \text{ L}}{8.314 \; \dfrac{L \cdot kPa}{mol \cdot K} \cdot 278.2 \text{ K}} = 0.698 \text{ mol}$

---

13. a.

$V = 1550 \text{ m}^3$

$P = 100{,}000$ Pa

$T = 0°C \rightarrow 0°C + 273.15 = 273.15$ K

$R = 8.314 \; \dfrac{J}{mol \cdot K}$

$PV = nRT \rightarrow n = \dfrac{PV}{RT} = \dfrac{100{,}000 \text{ Pa} \cdot 1550 \text{ m}^3}{8.314 \; \dfrac{J}{mol \cdot K} \cdot 273.15 \text{ K}} = 68{,}300 \text{ mol}$

---

13. b.

$68{,}300 \text{ mol CH}_4 \cdot \dfrac{16.043 \text{ g}}{mol} = 1{,}095{,}737 \text{ g} \cdot \dfrac{1 \text{ kg}}{1000 \text{ g}} = 1096 \text{ kg}$

When rounded to 3 sig digs, this gives $\boxed{1.10 \times 10^3 \text{ kg}}$

---

13. c.

$\rho = \dfrac{m}{V} = \dfrac{1096 \text{ kg}}{1550 \text{ m}^3} = 0.707 \; \dfrac{kg}{m^3}$

---

13. d.

$68{,}300 \text{ mol} \cdot \dfrac{6.022 \times 10^{23} \text{ particles}}{mol} = 4.11 \times 10^{28} \text{ particles}$

There is one carbon atom in each molecule of CH_4, giving a total of 4.11×10^{28} C atoms.

---

14. a.

$$V = 4960 \text{ m}^3 \cdot \frac{1000 \text{ L}}{1 \text{ m}^3} = 4,960,000 \text{ L}$$

$T = 17°C \rightarrow 17°C + 273.15 = 290.2 \text{ K}$

$P = 1.00 \text{ atm}$

$$R = 0.08206 \frac{\text{L} \cdot \text{atm}}{\text{mol} \cdot \text{K}}$$

$$PV = nRT \rightarrow n = \frac{PV}{RT} = \frac{1.00 \text{ atm} \cdot 4,960,000 \text{ L}}{0.08206 \dfrac{\text{L} \cdot \text{atm}}{\text{mol} \cdot \text{K}} \cdot 290.2 \text{ K}} = 208,300 \text{ mol}$$

$$208,300 \text{ mol He} \cdot \frac{4.003 \text{ g}}{\text{mol}} = 834,000 \text{ g} \cdot \frac{1 \text{ kg}}{1000 \text{ g}} = 834 \text{ kg}$$

---

14. b.

The number of moles would be the same. Hydrogen is a diatomic gas, so its molar mass is 2.015 g/mol.

$$208,300 \text{ mol H} \cdot \frac{2.0158 \text{ g}}{\text{mol}} = 419,900 \text{ g} \cdot \frac{1 \text{ kg}}{1000 \text{ g}} = 419.9 \text{ kg}$$

Rounding this value to 3 sig digs requires expressing it in scientific notation, giving 4.20×10^2 kg.

---

15. a.

$V = 2.000 \text{ L}$

$T = 27.0°C \rightarrow 27.0°C + 273.2 = 300.2 \text{ K}$

$P = 23.11 \text{ atm}$

$$R = 0.08206 \frac{\text{L} \cdot \text{atm}}{\text{mol} \cdot \text{K}}$$

$$PV = nRT \rightarrow n = \frac{PV}{RT} = \frac{23.11 \text{ atm} \cdot 2.000 \text{ L}}{0.08206 \dfrac{\text{L} \cdot \text{atm}}{\text{mol} \cdot \text{K}} \cdot 300.2 \text{ K}} = 1.876 \text{ mol}$$

---

15. b.

$$\frac{133.0 \text{ g}}{1.876 \text{ mol}} = 70.90 \frac{\text{g}}{\text{mol}}$$

All halogens are diatomic, which means the above molecular mass is twice the atomic mass. Dividing by 2 we get 35.45 g/mol, which is the atomic mass of chlorine.

---

16. a.

$V = 275$ L

$T = 29°C \rightarrow 29°C + 273.2 = 302.2$ K

$P = 765$ mm Hg $= 765$ Torr

$R = 62.36 \dfrac{\text{L} \cdot \text{Torr}}{\text{mol} \cdot \text{K}}$

$PV = nRT \rightarrow n = \dfrac{PV}{RT} = \dfrac{765 \text{ Torr} \cdot 275 \text{ L}}{62.36 \dfrac{\text{L} \cdot \text{Torr}}{\text{mol} \cdot \text{K}} \cdot 302.2 \text{ K}} = 11.16$ mol

Rounding this value to 2 sig digs gives 11.2 mol.

16. b.

At the new altitude, the number of moles of helium will be the same. We will use the value with the extra sig dig and round off to 3 sig digs at the end.

$n = 11.16$ mol

$T = 221$ K

$P = 25.0$ kPa

$R = 8.314 \dfrac{\text{L} \cdot \text{kPa}}{\text{mol} \cdot \text{K}}$

$PV = nRT \rightarrow V = \dfrac{nRT}{P} = \dfrac{11.16 \text{ mol} \cdot 8.314 \dfrac{\text{L} \cdot \text{kPa}}{\text{mol} \cdot \text{K}} \cdot 221 \text{ K}}{25.0 \text{ kPa}} = 820$ L

In order to show 3 sig digs, we write this value as 8.20×10^2 L.

17.

In this problem, only T and P and variables. Write the ideal gas law with these variables on one side and all the constants on the other side:

$PV = nRT \rightarrow \dfrac{P}{T} = \dfrac{nR}{V} = k$

Since P/T is a constant, P/T at one set of conditions is equal to P/T at any other set of conditions. So we can write

$\dfrac{P_1}{T_1} = \dfrac{P_2}{T_2}$

We must work only with absolute measurement values for P and T.

$P_{abs} = P_{gauge} + P_{atm}$

$P_1 = 35$ psig $+ 14.7$ psig $= 49.7$ psia

$T_1 = 36°C \rightarrow 36°C + 273.2 = 309.2$ K

$T_2 = -5°C \rightarrow -5°C + 273.2 = 268.2$ K

$\dfrac{P_1}{T_1} = \dfrac{P_2}{T_2} \rightarrow P_2 = P_1 \cdot \dfrac{T_2}{T_1} = 49.7 \text{ psia} \cdot \dfrac{268.2 \text{ K}}{309.2 \text{ K}} = 43.1$ psia

Finally, we must convert this result back to a gauge pressure with 2 sig digs.

$P_{abs} = P_{gauge} + P_{atm}$

$P_{gauge} = P_{abs} - P_{atm} = 43.1 \text{ psia} - 14.7 \text{ psi} = 28 \text{ psig}$

---

18. a.

As in the previous problem, the pressure must be converted to an absolute pressure.

$V = 12.00 \text{ L}$

$P = 119.0 \text{ bar (g)} \rightarrow 119.0 \text{ bar} + 1.01325 \text{ bar} = 120.01 \text{ bar (abs)} \cdot \dfrac{100 \text{ kPa}}{1 \text{ bar}} = 12{,}001 \text{ kPa}$

$T = 22.00°C \rightarrow 22.00°C + 273.15 = 295.15 \text{ K}$

$R = 8.314 \dfrac{\text{L} \cdot \text{kPa}}{\text{mol} \cdot \text{K}}$

$PV = nRT \rightarrow n = \dfrac{PV}{RT} = \dfrac{12{,}001 \text{ kPa} \cdot 12.00 \text{ L}}{8.314 \dfrac{\text{L} \cdot \text{kPa}}{\text{mol} \cdot \text{K}} \cdot 295.15 \text{ K}} = 58.688 \text{ mol}$

Rounding to 4 sig digs we have 58.69 mol.

---

18. b.

$n = 58.688 \text{ mol}$

$P = 100 \text{ kPa}$

$T = 0.0°C \rightarrow 0.0°C + 273.15 = 273.15 \text{ K}$

$R = 8.314 \dfrac{\text{L} \cdot \text{kPa}}{\text{mol} \cdot \text{K}}$

$PV = nRT \rightarrow V = \dfrac{nRT}{P} = \dfrac{58.688 \text{ mol} \cdot 8.314 \dfrac{\text{L} \cdot \text{kPa}}{\text{mol} \cdot \text{K}} \cdot 273.15 \text{ K}}{100 \text{ kPa}} = 1333 \text{ L}$

---

19. a.

The molar mass is the mass divided by the number of moles (g/mol).

$V = 0.975 \text{ L}$

$T = 26°C \rightarrow 26°C + 273.2 = 299.2 \text{ K}$

$P = 868 \text{ Torr}$

$R = 62.36 \dfrac{\text{L} \cdot \text{Torr}}{\text{mol} \cdot \text{K}}$

$PV = nRT \rightarrow n = \dfrac{PV}{RT} = \dfrac{868 \text{ Torr} \cdot 0.975 \text{ L}}{62.36 \dfrac{\text{L} \cdot \text{Torr}}{\text{mol} \cdot \text{K}} \cdot 299.2 \text{ K}} = 0.04535 \text{ mol}$

molar mass: $\dfrac{0.942 \text{ g}}{0.04535 \text{ mol}} = 20.8 \dfrac{\text{g}}{\text{mol}}$

19. b.

$$V = 888 \text{ mL} \cdot \frac{1 \text{ L}}{1000 \text{ mL}} = 0.888 \text{ L}$$

$$T = -32°C \rightarrow -32°C + 273.2 = 241.2 \text{ K}$$

$$P = 2.14 \text{ bar} \cdot \frac{100 \text{ kPa}}{1 \text{ bar}} = 214 \text{ kPa}$$

$$R = 8.314 \frac{\text{L} \cdot \text{kPa}}{\text{mol} \cdot \text{K}}$$

$$PV = nRT \rightarrow n = \frac{PV}{RT} = \frac{214 \text{ kPa} \cdot 0.888 \text{ L}}{8.314 \frac{\text{L} \cdot \text{kPa}}{\text{mol} \cdot \text{K}} \cdot 241.2 \text{ K}} = 0.09476 \text{ mol}$$

molar mass: $\dfrac{0.651 \text{ g}}{0.09476 \text{ mol}} = 6.87 \dfrac{\text{g}}{\text{mol}}$

20.

This is a challenging problem that illustrates a common solution strategy. Don't give up and look at the solution too soon! Focus on how to get from the given info to the required result.

Because of the high temperature, the hydrogen in stars exists as separate atoms rather than as H_2 molecules. Calculating the average molar mass:

$$0.670 \cdot 1.0079 \frac{\text{g}}{\text{mol}} + 0.330 \cdot 4.0026 \frac{\text{g}}{\text{mol}} = 0.675 \frac{\text{g}}{\text{mol}} + 1.32 \frac{\text{g}}{\text{mol}} = 1.996 \frac{\text{g}}{\text{mol}}$$

Next, the key to the problem is to notice that using the density equation we can express the mass as $\rho \cdot V$, and we can also express it as $(1.996 \text{ g/mol}) \cdot n$. Equating these gives us a value for V/n. We can then solve the ideal gas law for V/n, insert the expression we have, and solve for T. First we get the expression for V/n in units of L/mol:

$$\rho = 1.45 \frac{\text{g}}{\text{cm}^3}$$

$$\rho = \frac{m}{V} \rightarrow m = \rho V = 1.45 \frac{\text{g}}{\text{cm}^3} \cdot V$$

$$m = 1.996 \frac{\text{g}}{\text{mol}} \cdot n$$

Equating these expressions for m,

$$1.45 \frac{\text{g}}{\text{cm}^3} \cdot V = 1.996 \frac{\text{g}}{\text{mol}} \cdot n$$

$$\frac{V}{n} = \frac{1.996 \frac{\text{g}}{\text{mol}}}{1.45 \frac{\text{g}}{\text{cm}^3}} = 1.377 \frac{\text{cm}^3}{\text{mol}} \cdot \frac{1 \text{ L}}{1000 \text{ cm}^3} = 0.001377 \frac{\text{L}}{\text{mol}}$$

Now we turn to the ideal gas law:

$$\frac{V}{n} = 0.001377 \ \frac{L}{mol}$$

$$P = 1.29 \times 10^9 \ atm$$

$$R = 0.08206 \ \frac{L \cdot atm}{mol \cdot K}$$

$$PV = nRT \rightarrow T = \frac{PV}{nR} = \frac{P}{R} \cdot \frac{V}{n} = \frac{1.29 \times 10^9 \ atm}{0.08206 \ \dfrac{L \cdot atm}{mol \cdot K}} \cdot 0.001377 \ \frac{L}{mol} = 21{,}600{,}000 \ K$$

21.

The solution strategy for this problem is similar to that of the previous problem. First, we solve for an expression for density based on the conditions and the molar mass, symbolized by M:

$$\rho = \frac{m}{V}$$

$$m = M \cdot n$$

$$\rho = \frac{M \cdot n}{V}$$

$$PV = nRT \rightarrow \frac{n}{V} = \frac{P}{RT}$$

$$\rho = \frac{M \cdot n}{V} = \frac{M \cdot P}{RT}$$

Now we solve this for each of the given gases.

CO_2:

$$M = 44.010 \ \frac{g}{mol}$$

$$P = 100 \ kPa$$

$$T = 273.15 \ K$$

$$R = 8.314 \ \frac{L \cdot kPa}{mol \cdot K}$$

$$\rho = \frac{MP}{RT} = \frac{44.010 \ \dfrac{g}{mol} \cdot 100 \ kPa}{8.314 \ \dfrac{L \cdot kPa}{mol \cdot K} \cdot 273.15 \ K} = 1.938 \ \frac{g}{L}$$

O_2:

$$M = 31.999 \ \frac{g}{mol}$$

$$P = 100 \ kPa$$

$$T = 273.15 \ K$$

$$R = 8.314 \ \frac{L \cdot kPa}{mol \cdot K}$$

$$\rho = \frac{MP}{RT} = \frac{31.999 \ \frac{g}{mol} \cdot 100 \ kPa}{8.314 \ \frac{L \cdot kPa}{mol \cdot K} \cdot 273.15 \ K} = 1.409 \ \frac{g}{L}$$

N_2:

$$M = 28.013 \ \frac{g}{mol}$$

$$P = 100 \ kPa$$

$$T = 273.15 \ K$$

$$R = 8.314 \ \frac{L \cdot kPa}{mol \cdot K}$$

$$\rho = \frac{MP}{RT} = \frac{28.013 \ \frac{g}{mol} \cdot 100 \ kPa}{8.314 \ \frac{L \cdot kPa}{mol \cdot K} \cdot 273.15 \ K} = 1.234 \ \frac{g}{L}$$

CH_4:

$$M = 16.043 \ \frac{g}{mol}$$

$$P = 100 \ kPa$$

$$T = 273.15 \ K$$

$$R = 8.314 \ \frac{L \cdot kPa}{mol \cdot K}$$

$$\rho = \frac{MP}{RT} = \frac{16.043 \ \frac{g}{mol} \cdot 100 \ kPa}{8.314 \ \frac{L \cdot kPa}{mol \cdot K} \cdot 273.15 \ K} = 0.7064 \ \frac{g}{L}$$

25.

$X_{N_2} = 0.820$

$X_{Ar} = 0.120$

$X_{CH_4} = 0.060$

$P_T = 1.61$ atm

$P_{N_2} = X_{N_2} \cdot P_T = 0.820 \cdot 1.61 \text{ atm} = 1.32 \text{ atm}$

$P_{Ar} = X_{Ar} \cdot P_T = 0.120 \cdot 1.61 \text{ atm} = 0.193 \text{ atm}$

$P_{CH_4} = X_{CH_4} \cdot P_T = 0.060 \cdot 1.61 \text{ atm} = 0.097 \text{ atm}$

---

26. a.

$X_{Ar} = 0.805$

$X_{O_2} = 0.180$

$X_{CO_2} = 0.015$

$P_T = 748$ Torr

$P_{Ar} = X_{Ar} \cdot P_T = 0.805 \cdot 748 \text{ Torr} = 602.1 \text{ Torr}$

For 3 sig digs, this rounds to 602 Torr.

---

26. b.

$P = 602.1$ Torr

$V = 165$ L

$T = 295$ K

$R = 62.36 \dfrac{\text{L} \cdot \text{Torr}}{\text{mol} \cdot \text{K}}$

$PV = nRT \rightarrow n = \dfrac{PV}{RT} = \dfrac{602.1 \text{ Torr} \cdot 165 \text{ L}}{62.36 \dfrac{\text{L} \cdot \text{Torr}}{\text{mol} \cdot \text{K}} \cdot 295 \text{ K}} = 5.40 \text{ mol}$

27.

The partial pressure of propane is given. From this and the number of moles of each gas the total pressure may be determined:

$$P_{C_3H_8} = 2.55 \text{ bar}$$

$$m_{C_3H_8} = 2.55 \text{ g}$$

$$m_{CH_4} = 1.01 \text{ g}$$

$$n_{C_3H_8} = 2.55 \text{ g C}_3\text{H}_8 \cdot \frac{\text{mol}}{44.096 \text{ g}} = 0.05783 \text{ mol}$$

$$n_{CH_4} = 1.01 \text{ g CH}_4 \cdot \frac{\text{mol}}{16.043 \text{ g}} = 0.06296 \text{ mol}$$

$$n_T = n_{C_3H_8} + n_{CH_4} = 0.05783 \text{ mol} + 0.06296 \text{ mol} = 0.1208 \text{ mol}$$

$$P_{C_3H_8} = \frac{n_{C_3H_8}}{n_T} \cdot P_T \rightarrow P_T = P_{C_3H_8} \cdot \frac{n_T}{n_{C_3H_8}} = 2.55 \text{ bar} \cdot \frac{0.1208 \text{ mol}}{0.05783 \text{ mol}} = 5.327 \text{ bar}$$

Rounding to 3 sig digs we have $P_T = 5.33$ bar. With the total pressure we can determine the partial pressure of methane:

$$P_{CH_4} = \frac{n_{CH_4}}{n_T} \cdot P_T = \frac{0.06296 \text{ mol}}{0.1208 \text{ mol}} \cdot 5.327 \text{ bar} = 2.78 \text{ bar}$$

Finally, the mole fractions are:

$$X_{C_3H_8} = \frac{n_{C_3H_8}}{n_T} = \frac{0.05783 \text{ mol}}{0.1208 \text{ mol}} = 0.479$$

$$X_{CH_4} = \frac{n_{CH_4}}{n_T} = \frac{0.06296 \text{ mol}}{0.1208 \text{ mol}} = 0.521$$

28. a.

O_2:

$V = 2.00$ L

$P = 1.55$ atm

$T = 23°C \rightarrow 23°C + 273.2 = 296.2$ K

$R = 0.08206 \dfrac{\text{L} \cdot \text{atm}}{\text{mol} \cdot \text{K}}$

$$PV = nRT \rightarrow n = \dfrac{PV}{RT} = \dfrac{1.55 \text{ atm} \cdot 2.00 \text{ L}}{0.08206 \dfrac{\text{L} \cdot \text{atm}}{\text{mol} \cdot \text{K}} \cdot 296.2 \text{ K}} = 0.1275 \text{ mol}$$

Rounding this value to 3 sig digs gives 0.128 mol.

H_2:

$V = 1.00$ L

$P = 1.05$ atm

$T = 23°C \rightarrow 23°C + 273.2 = 296.2$ K

$R = 0.08206 \dfrac{\text{L} \cdot \text{atm}}{\text{mol} \cdot \text{K}}$

$$PV = nRT \rightarrow n = \dfrac{PV}{RT} = \dfrac{1.05 \text{ atm} \cdot 1.00 \text{ L}}{0.08206 \dfrac{\text{L} \cdot \text{atm}}{\text{mol} \cdot \text{K}} \cdot 296.2 \text{ K}} = 0.04320 \text{ mol}$$

Rounding this value to 3 sig digs gives 0.0432 mol.

28. b.

$V = 3.00$ L

$n_{O_2} = 0.1275$ mol

$T = 23°C \rightarrow 23°C + 273.2 = 296.2$ K

$R = 0.08206 \dfrac{\text{L} \cdot \text{atm}}{\text{mol} \cdot \text{K}}$

$$PV = nRT \rightarrow P_{O_2} = \dfrac{n_{O_2} RT}{V} = \dfrac{0.1275 \text{ mol} \cdot 0.08206 \dfrac{\text{L} \cdot \text{atm}}{\text{mol} \cdot \text{K}} \cdot 296.2 \text{ K}}{3.00 \text{ L}} = 1.033 \text{ atm}$$

Rounding this value to 3 sig digs gives 1.03 atm.

$V = 3.00$ L

$n_{H_2} = 0.04320$ mol

$T = 23°C \rightarrow 23°C + 273.2 = 296.2$ K

$R = 0.08206 \dfrac{\text{L} \cdot \text{atm}}{\text{mol} \cdot \text{K}}$

$$PV = nRT \rightarrow P_{H_2} = \dfrac{n_{H_2} RT}{V} = \dfrac{0.04320 \text{ mol} \cdot 0.08206 \dfrac{\text{L} \cdot \text{atm}}{\text{mol} \cdot \text{K}} \cdot 296.2 \text{ K}}{3.00 \text{ L}} = 0.3500 \text{ atm}$$

Rounding this value to 3 sig digs gives 0.350 atm.

28. c.

$P_T = P_{O_2} + P_{H_2} = 1.033 \text{ atm} + 0.3500 \text{ atm} = 1.38 \text{ atm}$

---

29. a/b.

First, determine the partial pressure of water at this temperature, which is its vapor pressure. From Table B.4, this is 2.6453 kPa.

$P_{water} = 2.6453 \text{ kPa} \cdot \dfrac{1 \text{ atm}}{101.325 \text{ kPa}} = 0.02611 \text{ atm}$

$P_T = 1.00 \text{ atm}$

$P_T = P_{water} + P_{N_2} \to P_{N_2} = P_T - P_{water} = 1.00 \text{ atm} - 0.02611 \text{ atm} = 0.974 \text{ atm}$

The addition rule requires us to round this partial pressure to the second decimal place, giving 0.97 atm. Now we calculate the number of moles using the ideal gas equation:

$P_{N_2} = 0.974 \text{ atm}$

$V = 285 \text{ mL} \cdot \dfrac{1 \text{ L}}{1000 \text{ mL}} = 0.285 \text{ L}$

$T = 22°C \to 22°C + 273.2 \text{ K} = 295.2 \text{ K}$

$R = 0.08206 \dfrac{\text{L} \cdot \text{atm}}{\text{mol} \cdot \text{K}}$

$PV = nRT \to n_{N_2} = \dfrac{P_{N_2} V}{RT} = \dfrac{0.974 \text{ atm} \cdot 0.285 \text{ L}}{0.08206 \dfrac{\text{L} \cdot \text{atm}}{\text{mol} \cdot \text{K}} \cdot 295.2 \text{ K}} = 0.01146 \text{ mol}$

With 2 sig digs, this is 0.011 mol. To obtain the mole fraction, we have to first obtain the number of moles of water vapor:

$P_{water} = 0.02611 \text{ atm}$

$V = 0.285 \text{ L}$

$T = 295.2 \text{ K}$

$R = 0.08206 \dfrac{\text{L} \cdot \text{atm}}{\text{mol} \cdot \text{K}}$

$PV = nRT \to n_{water} = \dfrac{P_{water} V}{RT} = \dfrac{0.02611 \text{ atm} \cdot 0.285 \text{ L}}{0.08206 \dfrac{\text{L} \cdot \text{atm}}{\text{mol} \cdot \text{K}} \cdot 295.2 \text{ K}} = 0.0003072 \text{ mol}$

This value still has 3 sig digs, so it rounds to 0.000307 mol. Now we can calculate the total number of moles and the mole fractions.

$n_T = n_{N_2} + n_{water} = 0.01146 \text{ mol} + 0.0003072 \text{ mol} = 0.01177 \text{ mol}$

$X_{N_2} = \dfrac{n_{N_2}}{n_T} = \dfrac{0.01146 \text{ mol}}{0.01177 \text{ mol}} = 0.974$

$X_{water} = \dfrac{n_{water}}{n_T} = \dfrac{0.0003072 \text{ mol}}{0.01177 \text{ mol}} = 0.0261$

The value for the moles of N_2 limits precision to the third decimal place in n_T, which in turn limits the precision of these mole fractions to 2 sig digs, giving 0.97 and 0.026.

30.

We need to use the partial pressure of H_2 to determine the number of moles of H_2. From this we can do the stoichiometry to determine the amount of zinc consumed. From Table B.4, the vapor pressure of water at these conditions is 28.38 Torr.

$$P_T = P_{H_2} + P_{water} \rightarrow P_{H_2} = P_T - P_{water} = 757 \text{ Torr} - 28.38 \text{ Torr} = 728.6 \text{ Torr}$$

$$V = 184 \text{ mL} \cdot \frac{1 \text{ L}}{1000 \text{ mL}} = 0.184 \text{ L}$$

$$T = 28°C \rightarrow 28°C + 273.2 = 301.2 \text{ K}$$

$$R = 62.36 \frac{\text{L} \cdot \text{Torr}}{\text{mol} \cdot \text{K}}$$

$$PV = nRT \rightarrow n_{H_2} = \frac{P_{H_2} V}{RT} = \frac{728.6 \text{ Torr} \cdot 0.184 \text{ L}}{62.36 \dfrac{\text{L} \cdot \text{Torr}}{\text{mol} \cdot \text{K}} \cdot 301.2 \text{ K}} = 0.007138 \text{ mol}$$

$$0.007138 \text{ mol } H_2 \cdot \frac{1 \text{ mol Zn}}{1 \text{ mol } H_2} = 0.007138 \text{ mol Zn}$$

$$0.007138 \text{ mol Zn} \cdot \frac{65.39 \text{ g}}{\text{mol}} = 0.467 \text{ g}$$

31.

$$2HgO \rightarrow 2Hg + O_2$$

From Table B-4, the vapor pressure of water at these conditions is 15.49 Torr.

$$P_T = P_{O_2} + P_{water} \rightarrow P_{O_2} = P_T - P_{water} = 725 \text{ Torr} - 15.49 \text{ Torr} = 709.5 \text{ Torr}$$

$$V = 355 \text{ mL} \cdot \frac{1 \text{ L}}{1000 \text{ mL}} = 0.355 \text{ L}$$

$$T = 18°C \rightarrow 18°C + 273.2 = 291.2 \text{ K}$$

$$R = 62.36 \frac{\text{L} \cdot \text{Torr}}{\text{mol} \cdot \text{K}}$$

$$PV = nRT \rightarrow n_{O_2} = \frac{P_{O_2} V}{RT} = \frac{709.5 \text{ Torr} \cdot 0.355 \text{ L}}{62.36 \dfrac{\text{L} \cdot \text{Torr}}{\text{mol} \cdot \text{K}} \cdot 291.2 \text{ K}} = 0.01387 \text{ mol}$$

$$0.01387 \text{ mol } O_2 \cdot \frac{2 \text{ mol HgO}}{1 \text{ mol } O_2} = 0.02774 \text{ mol HgO}$$

$$0.02774 \text{ mol HgO} \cdot \frac{216.6 \text{ g}}{\text{mol}} = 6.01 \text{ g}$$

32.

$$2CO + O_2 \rightarrow 2CO_2$$

$$5.0 \text{ L CO} \cdot \frac{1 \text{ mol } O_2}{2 \text{ mol CO}} = 2.5 \text{ L } O_2$$

$$5.0 \text{ L CO} \cdot \frac{2 \text{ mol } CO_2}{2 \text{ mol CO}} = 5.0 \text{ L } CO_2$$

33.

$$C_3H_8 + 5O_2 \rightarrow 3CO_2 + 4H_2O$$

$$30.00 \text{ m}^3 \text{ C}_3H_8 \cdot \frac{5 \text{ mol O}_2}{1 \text{ mol C}_3H_8} = 150.0 \text{ m}^3 \text{ O}_2$$

$$30.00 \text{ m}^3 \text{ C}_3H_8 \cdot \frac{3 \text{ mol CO}_2}{1 \text{ mol C}_3H_8} = 90.00 \text{ m}^3 \text{ CO}_2$$

$$30.00 \text{ m}^3 \text{ C}_3H_8 \cdot \frac{4 \text{ mol H}_2O}{1 \text{ mol C}_3H_8} = 120.0 \text{ m}^3 \text{ H}_2O$$

34.

$$2Fe(OH)_3 \rightarrow Fe_2O_3 + 3H_2O$$

$$V = 2.00 \text{ L}$$

$$P = 101.325 \text{ kPa}$$

$$T = 390 \text{ K}$$

$$R = 8.314 \frac{\text{L} \cdot \text{kPa}}{\text{mol} \cdot \text{K}}$$

$$PV = nRT \rightarrow n = \frac{PV}{RT} = \frac{101.325 \text{ kPa} \cdot 2.00 \text{ L}}{8.314 \dfrac{\text{L} \cdot \text{kPa}}{\text{mol} \cdot \text{K}} \cdot 390 \text{ K}} = 0.0625 \text{ mol H}_2O$$

$$0.0625 \text{ mol H}_2O \cdot \frac{2 \text{ mol Fe(OH)}_3}{3 \text{ mol H}_2O} = 0.0417 \text{ mol Fe(OH)}_3$$

$$0.0417 \text{ mol Fe(OH)}_3 \cdot \frac{106.9 \text{ g}}{\text{mol}} = 4.5 \text{ g Fe(OH)}_3$$

$$0.0625 \text{ mol H}_2O \cdot \frac{1 \text{ mol Fe}_2O_3}{3 \text{ mol H}_2O} = 0.0208 \text{ mol Fe}_2O_3$$

$$0.0208 \text{ mol Fe}_2O_3 \cdot \frac{159.7 \text{ g}}{\text{mol}} = 3.3 \text{ g Fe}_2O_3$$

35.

$V = 31,150 \text{ L}$

$P = 101.325 \text{ kPa}$

$T = 24°C \rightarrow 24°C + 273.2 = 297.2 \text{ K}$

$R = 8.314 \dfrac{\text{L} \cdot \text{kPa}}{\text{mol} \cdot \text{K}}$

$PV = nRT \rightarrow n = \dfrac{PV}{RT} = \dfrac{101.325 \text{ kPa} \cdot 31,150 \text{ L}}{8.314 \dfrac{\text{L} \cdot \text{kPa}}{\text{mol} \cdot \text{K}} \cdot 297.2 \text{ K}} = 1277 \text{ mol H}_2$

$1277 \text{ mol H}_2 \cdot \dfrac{1 \text{ mol Fe}}{1 \text{ mol H}_2} = 1277 \text{ mol Fe}$

$1277 \text{ mol Fe} \cdot \dfrac{55.847 \text{ g}}{\text{mol}} = 71,400 \text{ g Fe} \cdot \dfrac{1 \text{ kg}}{1000 \text{ g}} = 71.4 \text{ kg}$

Note that although the temperature was given with 2 sig digs, this value has 3 sig digs after being converted to kelvins. Thus, the result has 3 sig digs.

36.

$\text{Fe} + \text{H}_2\text{SO}_4 \rightarrow \text{FeSO}_4 + \text{H}_2$

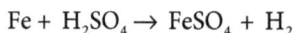
$1277 \text{ mol H}_2 \cdot \dfrac{1 \text{ mol Fe}}{1 \text{ mol H}_2} = 1277 \text{ mol Fe}$

$1277 \text{ mol Fe} \cdot \dfrac{55.847 \text{ g}}{\text{mol}} = 71,400 \text{ g Fe} \cdot \dfrac{1 \text{ kg}}{1000 \text{ g}} = 71.4 \text{ kg}$

37.

$\text{Mg} + 2\text{HCl} \rightarrow \text{MgCl}_2 + \text{H}_2$

$V = 0.015 \text{ L}$

$P = 101.325 \text{ kPa}$

$T = 273.15 \text{ K}$

$R = 8.314 \dfrac{\text{L} \cdot \text{kPa}}{\text{mol} \cdot \text{K}}$

$PV = nRT \rightarrow n = \dfrac{PV}{RT} = \dfrac{101.325 \text{ kPa} \cdot 0.015 \text{ L}}{8.314 \dfrac{\text{L} \cdot \text{kPa}}{\text{mol} \cdot \text{K}} \cdot 273.15 \text{ K}} = 6.69 \times 10^{-4} \text{ mol H}_2$

$6.69 \times 10^{-4} \text{ mol H}_2 \cdot \dfrac{1 \text{ mol Mg}}{1 \text{ mol H}_2} = 6.69 \times 10^{-4} \text{ mol Mg}$

$6.69 \times 10^{-4} \text{ mol Mg} \cdot \dfrac{24.305 \text{ g}}{\text{mol}} = 0.016 \text{ g}$

38.

From Table B.4, the vapor pressure of water at these conditions is 17.55 Torr.

$$0.883 \text{ g CaC}_2 \cdot \frac{\text{mol}}{64.10 \text{ g}} = 0.01378 \text{ mol CaC}_2$$

$$P_T = P_{C_2H_2} + P_{water} \rightarrow P_{C_2H_2} = P_T - P_{water} = 735 \text{ Torr} - 17.55 \text{ Torr} = 717.5 \text{ Torr}$$

$$T = 20°C \rightarrow 20°C + 273.2 = 293.2 \text{ K}$$

$$R = 62.36 \frac{\text{L} \cdot \text{Torr}}{\text{mol} \cdot \text{K}}$$

$$PV = nRT \rightarrow V_{C_2H_2} = \frac{n_{C_2H_2}RT}{P_{C_2H_2}} = \frac{0.01378 \text{ mol} \cdot 62.36 \dfrac{\text{L} \cdot \text{Torr}}{\text{mol} \cdot \text{K}} \cdot 293.2 \text{ K}}{717.5 \text{ Torr}} = 0.351 \text{ L}$$

39.

$$2KI + Cl_2 \rightarrow 2KCl + I_2$$

39. a.

$$V = 7.75 \text{ L}$$

$$P = 100 \text{ kPa}$$

$$T = 273.15 \text{ K}$$

$$R = 8.314 \frac{\text{L} \cdot \text{kPa}}{\text{mol} \cdot \text{K}}$$

$$PV = nRT \rightarrow n = \frac{PV}{RT} = \frac{100 \text{ kPa} \cdot 7.75 \text{ L}}{8.314 \dfrac{\text{L} \cdot \text{kPa}}{\text{mol} \cdot \text{K}} \cdot 273.15 \text{ K}} = 0.3413 \text{ mol } I_2$$

$$0.3413 \text{ mol } I_2 \cdot \frac{2 \text{ mol KI}}{1 \text{ mol } I_2} = 0.6826 \text{ mol KI}$$

$$0.3413 \text{ mol } I_2 \cdot \frac{1 \text{ mol Cl}_2}{1 \text{ mol } I_2} = 0.3413 \text{ mol Cl}_2$$

$$0.3413 \text{ mol } I_2 \cdot \frac{2 \text{ mol KCl}}{1 \text{ mol } I_2} = 0.6826 \text{ mol KCl}$$

Rounding these to 3 sig digs gives 0.341 mol I_2, 0.683 mol KI, 0.341 mol Cl_2, and 0.683 mol KCl.

---

39. b.

$$0.3413 \text{ mol } I_2 \cdot \frac{253.8 \text{ g}}{\text{mol}} = 86.6 \text{ g } I_2$$

$$0.6826 \text{ mol KI} \cdot \frac{165.2 \text{ g}}{\text{mol}} = 113 \text{ g KI}$$

$$0.3413 \text{ mol Cl}_2 \cdot \frac{70.91 \text{ g}}{\text{mol}} = 24.2 \text{ g Cl}_2$$

$$0.6286 \text{ mol KCl} \cdot \frac{74.55 \text{ g}}{\text{mol}} = 50.9 \text{ g KCl}$$

40.

$2Al + 6HCl \rightarrow 2AlCl_3 + 3H_2$

40. a.

$110.0 \text{ g HCl} \cdot \dfrac{\text{mol}}{36.4606 \text{ g}} = 3.017 \text{ mol HCl}$

$25.00 \text{ g Al} \cdot \dfrac{\text{mol}}{26.9815 \text{ g}} = 0.92656 \text{ mol Al}$

$0.92656 \text{ mol Al} \cdot \dfrac{6 \text{ mol HCl}}{2 \text{ mol Al}} = 2.78 \text{ mol HCl}$

More HCl than this is available, so Al is the limiting reactant.

40. b.

$0.92656 \text{ mol Al} \cdot \dfrac{3 \text{ mol H}_2}{2 \text{ mol Al}} = 1.3898 \text{ mol H}_2$

$P = 101.325 \text{ kPa}$

$T = 32.0°C \rightarrow 32.0°C + 273.15 = 305.15 \text{ K}$

$R = 8.314 \dfrac{\text{L} \cdot \text{kPa}}{\text{mol} \cdot \text{K}}$

$PV = nRT \rightarrow V = \dfrac{nRT}{P} = \dfrac{1.3898 \text{ mol} \cdot 8.314 \dfrac{\text{L} \cdot \text{kPa}}{\text{mol} \cdot \text{K}} \cdot 305.15 \text{ K}}{101.325 \text{ kPa}} = 34.80 \text{ L}$

41.

$CO + 2H_2 \rightarrow CH_3OH$

41. a.

$660.0 \text{ m}^3 \text{ CO} \cdot \dfrac{2 \text{ m}^3 \text{ H}_2}{1 \text{ m}^3 \text{ CO}} = 1320 \text{ m}^3 \text{ H}_2$

This much H_2 is not available. Thus, H_2 is the limiting reactant and CO is present in excess.

41. b.

$1210.0 \text{ m}^3 \text{ H}_2 \cdot \dfrac{1 \text{ m}^3 \text{ CO}}{2 \text{ m}^3 \text{ H}_2} = 605.0 \text{ m}^3 \text{ CO}$

$660.0 \text{ m}^3 - 605.0 \text{ m}^3 = 55.0 \text{ m}^3$

41. c.

$1210.0 \text{ m}^3 \text{ H}_2 \cdot \dfrac{1 \text{ m}^3 \text{ CH}_3OH}{2 \text{ m}^3 \text{ H}_2} = 605.0 \text{ m}^3 \text{ CH}_3OH$

42.

$4C_3H_5N_3O_9 \rightarrow 10H_2O + 12CO_2 + 6N_2 + O_2$

42. a.

$$1.00 \text{ g H}_2\text{O} \cdot \frac{\text{mol}}{18.015 \text{ g}} = 0.05551 \text{ mol H}_2\text{O}$$

$$0.05551 \text{ mol H}_2\text{O} \cdot \frac{12 \text{ mol CO}_2}{10 \text{ mol H}_2\text{O}} = 0.06661 \text{ mol CO}_2$$

$$0.05551 \text{ mol H}_2\text{O} \cdot \frac{6 \text{ mol N}_2}{10 \text{ mol H}_2\text{O}} = 0.03331 \text{ mol N}_2$$

$$0.05551 \text{ mol H}_2\text{O} \cdot \frac{1 \text{ mol O}_2}{10 \text{ mol H}_2\text{O}} = 0.005551 \text{ mol O}_2$$

$$n_T = 0.05551 \text{ mol} + 0.06661 \text{ mol} + 0.03331 \text{ mol} + 0.00555 \text{ mol} = 0.16098 \text{ mol}$$

$$P = 50.0 \text{ bar} \cdot \frac{100 \text{ kPa}}{1 \text{ bar}} = 5000.0 \text{ kPa}$$

$$T = 350°\text{C} \rightarrow 350°\text{C} + 273 = 623 \text{ K}$$

$$R = 8.314 \frac{\text{L} \cdot \text{kPa}}{\text{mol} \cdot \text{K}}$$

$$PV = nRT \rightarrow V = \frac{nRT}{P} = \frac{0.16098 \text{ mol} \cdot 8.314 \dfrac{\text{L} \cdot \text{kPa}}{\text{mol} \cdot \text{K}} \cdot 623 \text{ K}}{5000.0 \text{ kPa}} = 0.17 \text{ L}$$

42. b.

$$0.05551 \text{ mol H}_2\text{O} \cdot \frac{4 \text{ mol C}_3\text{H}_5\text{N}_3\text{O}_9}{10 \text{ mol H}_2\text{O}} = 0.02220 \text{ mol C}_3\text{H}_5\text{N}_3\text{O}_9$$

$$0.02220 \text{ mol C}_3\text{H}_5\text{N}_3\text{O}_9 \cdot \frac{227.1 \text{ g}}{\text{mol}} = 5.0 \text{ g}$$

43.

$$CO_2 : M = 44.01 \frac{\text{g}}{\text{mol}}$$

$$CCl_4 : M = 153.8 \frac{\text{g}}{\text{mol}}$$

$$CO_2 \text{ effusion rate} = 6.1 \times 10^{-2} \frac{\text{mol}}{\text{s}}$$

$$\frac{CO_2 \text{ effusion rate}}{CCl_4 \text{ effusion rate}} = \frac{\sqrt{M_{CCl_4}}}{\sqrt{M_{CO_2}}}$$

$$CCl_4 \text{ effusion rate} = CO_2 \text{ effusion rate} \cdot \frac{\sqrt{M_{CO_2}}}{\sqrt{M_{CCl_4}}} = 6.1 \times 10^{-2} \frac{\text{mol}}{\text{s}} \cdot \frac{\sqrt{44.01 \dfrac{\text{g}}{\text{mol}}}}{\sqrt{153.8 \dfrac{\text{g}}{\text{mol}}}} = 0.033 \frac{\text{mol}}{\text{s}}$$

44.

$$Cl_2 : M = 70.91 \frac{g}{mol}$$

$$SO_2 : M = 64.06 \frac{g}{mol}$$

$$v_{Cl_2} = 0.0421 \frac{m}{s}$$

$$\frac{v_{Cl_2}}{v_{SO_2}} = \frac{\sqrt{M_{SO_2}}}{\sqrt{M_{Cl_2}}}$$

$$v_{SO_2} = v_{Cl_2} \cdot \frac{\sqrt{M_{Cl_2}}}{\sqrt{M_{SO_2}}} = 0.0421 \frac{m}{s} \cdot \sqrt{\frac{70.91 \frac{g}{mol}}{64.06 \frac{g}{mol}}} = 0.0443 \frac{m}{s}$$

45.

$$\frac{\text{He effusion rate}}{\text{X effusion rate}} = 7.1 = \frac{\sqrt{M_X}}{\sqrt{M_{He}}}$$

$$\sqrt{M_X} = 7.1 \cdot \sqrt{M_{He}} = 7.1 \cdot \sqrt{4.0026 \frac{g}{mol}}$$

$$M_X = 7.1^2 \cdot 4.0026 \frac{g}{mol} = 201 \frac{g}{mol}$$

Writing this result with two 2 sig digs, we have 2.0×10^2 g/mol.

46.

$$\frac{M_A}{M_B} = 3.4$$

$$\frac{\text{gas A effusion rate}}{\text{gas B effusion rate}} = \frac{\sqrt{M_B}}{\sqrt{M_A}} = \sqrt{\frac{M_B}{M_A}} = \frac{1}{\sqrt{\frac{M_A}{M_B}}} = \frac{1}{\sqrt{3.4}} = \frac{1}{1.8} \approx 0.54$$

47.

$$E = 3.20 \times 10^{-18} \text{ J}$$

$$E = \frac{hv}{\lambda} \rightarrow \lambda = \frac{hv}{E} = \frac{6.626 \times 10^{-34} \text{ J} \cdot \text{s} \cdot 2.9979 \times 10^8 \frac{m}{s}}{3.20 \times 10^{-18} \text{ J}} = 6.21 \times 10^{-8} \text{ m} \cdot \frac{1 \times 10^9 \text{ nm}}{1 \text{ m}} = 62.1 \text{ nm}$$

This wavelength is in the UV region.

48.

$$\frac{3 \cdot 12.011}{227.087} = 0.15867 \rightarrow 15.867\% \text{ C}$$

$$\frac{5 \cdot 1.0079}{227.087} = 0.022192 \rightarrow 2.2192\% \text{ H}$$

$$\frac{3 \cdot 14.0067}{227.087} = 0.185040 \rightarrow 18.5040\% \text{ N}$$

$$\frac{9 \cdot 15.9994}{227.087} = 0.634094 \rightarrow 63.4094\% \text{ O}$$

49. a.

$$53.5 \text{ kg} \cdot \frac{1000 \text{ g}}{1 \text{ kg}} = 53,500 \text{ g Fe}$$

$$53,500 \text{ g Fe} \cdot \frac{\text{mol}}{55.847 \text{ g}} = 958.0 \text{ mol Fe}$$

$$958.0 \text{ mol Fe} \cdot \frac{1 \text{ mol H}_2}{1 \text{ mol Fe}} = 958.0 \text{ mol H}_2$$

$$V = 19,625 \text{ L}$$

$$P = 101.325 \text{ kPa}$$

$$T = 24°\text{C} \rightarrow 24°\text{C} + 273.2 = 297.2 \text{ K}$$

$$R = 8.314 \frac{\text{L} \cdot \text{kPa}}{\text{mol} \cdot \text{K}}$$

$$PV = nRT \rightarrow n = \frac{PV}{RT} = \frac{101.325 \text{ kPa} \cdot 19,625 \text{ L}}{8.314 \frac{\text{L} \cdot \text{kPa}}{\text{mol} \cdot \text{K}} \cdot 297.2 \text{ K}} = 804.7 \text{ mol}$$

$$\frac{804.7 \text{ mol}}{958.0 \text{ mol}} = 84.0\%$$

49. b.

$$804.7 \text{ mol} \cdot \frac{6.022 \times 10^{23} \text{ particles}}{\text{mol}} = 4.85 \times 10^{26} \text{ particles (molecules)}$$

49. c.

$$958.0 \text{ mol Fe} \cdot \frac{1 \text{ mol FeCl}_2}{1 \text{ mol Fe}} = 958.0 \text{ mol FeCl}_2$$

$$958.0 \text{ mol FeCl}_2 \cdot \frac{126.8 \text{ g}}{\text{mol}} = 121,000 \text{ g} \cdot \frac{1 \text{ kg}}{1000 \text{ g}} = 121 \text{ kg}$$

51.

$$C_{ice} = 0.0364 \ \frac{kJ}{mol \cdot K}$$

$$C_{water} = 0.0752 \ \frac{kJ}{mol \cdot K}$$

$$H_f = 6.01 \ \frac{kJ}{mol}$$

$$m = 22.0 \ kg \cdot \frac{1000 \ g}{1 \ kg} = 22,000 \ g$$

$$n = 22,000 \ g \cdot \frac{mol}{18.015 \ g} = 1221 \ mol$$

$$\Delta T = T_f - T_i = 20.0°C - 0.0°C = 20.0°C = 20.0 \ K$$

$$Q_{melt} = nH_f = 1221 \ mol \cdot 6.01 \ \frac{kJ}{mol} = 7340 \ kJ$$

$$Q_{warm} = Cn\Delta T = 0.0752 \ \frac{kJ}{mol \cdot K} \cdot 1221 \ mol \cdot 20.0 \ K = 1840 \ kJ$$

$$Q_{total} = Q_{melt} + Q_{warm} = 7340 \ kJ + 1840 \ kJ = 9180 \ kJ$$

Chapter 8

---

20. d.

$$\frac{0.7-0.54}{0.7}\cdot 100\% = 23\%$$

---

21. d.

$$Al_2(SO_4)_3 \rightarrow 342.2\ \frac{g}{mol}$$

$$0.282\cdot 45\ g = 12.7\ g\ Al_2(SO_4)_3$$

$$12.7\ g\cdot\frac{1\ mol}{342.2\ g} = 0.0371\ mol$$

$$H_2O \rightarrow 18.02\ \frac{g}{mol}$$

$$0.718\cdot 45\ g = 32.3\ g\ H_2O$$

$$32.3\ g\cdot\frac{1\ mol}{18.02\ g} = 1.79\ mol$$

total moles $= 1.79 + 0.04 = 1.83$

$$Al_2(SO_4)_3:\ \frac{0.0371}{1.83\ mol} = 0.02$$

$$H_2O:\ \frac{1.79\ mol}{1.83\ mol} = 0.98$$

---

22. a.

$$8.00\ \frac{mol}{L}\cdot 2.25\ L = 18.0\ mol$$

$$18.0\ mol\ NaOH\cdot\frac{40.00\ g}{mol} = 719.9\ g$$

To round this to 3 sig digs, write it in scientific notation: 7.20×10^2 g.

---

22. b.

$$15.0\ \frac{mol}{L}\cdot 1.00\ L = 15.0\ mol$$

$$15.0\ mol\ CH_3COOH\cdot\frac{60.05\ g}{mol} = 901\ g$$

23.

$$65.11 \text{ g } (NH_4)_2 SO_4 \cdot \frac{\text{mol}}{132.140 \text{ g}} = 0.4927 \text{ mol}$$

$$125 \text{ mL} \cdot \frac{1 \text{ L}}{1000 \text{ mL}} = 0.125 \text{ L}$$

$$\frac{0.4927 \text{ mol}}{0.125 \text{ L}} = 3.94 \text{ M}$$

24.

$$1.25 \frac{\text{mol}}{\text{L}} \cdot \frac{164.1 \text{ g}}{\text{mol}} = 205.1 \frac{\text{g}}{\text{L}}$$

$$205.1 \frac{\text{g}}{\text{L}} \cdot 3.10 \text{ L} = 636 \text{ g}$$

25.

$$1.2 \frac{\text{mol}}{\text{L}} \cdot \frac{169.87 \text{ g}}{\text{mol}} = 203.8 \frac{\text{g}}{\text{L}}$$

$$22.5 \text{ g} \cdot \frac{1 \text{ L}}{203.8 \text{ g}} = 0.11 \text{ L} \cdot \frac{1000 \text{ mL}}{1 \text{ L}} = 110 \text{ mL}$$

26.

$$CuSO_4 + Fe \rightarrow FeSO_4 + Cu$$

$$7.7 \text{ g Cu} \cdot \frac{\text{mol}}{63.546 \text{ g}} = 0.1212 \text{ mol}$$

$$0.1212 \text{ mol} \cdot \frac{1 \text{ mol CuSO}_4}{1 \text{ mol Cu}} = 0.1212 \text{ mol CuSO}_4$$

$$54.0 \text{ mL} \cdot \frac{1 \text{ L}}{1000 \text{ mL}} = 0.0540 \text{ L}$$

$$\frac{0.1212 \text{ mol}}{0.0540 \text{ L}} = 2.2 \text{ M}$$

27. a.

$$Na_2SO_4(aq) + Ba(NO_3)_2(aq) \rightarrow 2NaNO_3(aq) + BaSO_4(s)$$

27. b.

$$475 \text{ mL} \cdot \frac{1 \text{ L}}{1000 \text{ mL}} = 0.475 \text{ L}$$

$$0.475 \text{ L} \cdot \frac{0.60 \text{ mol}}{\text{L}} = 0.285 \text{ mol Na}_2SO_4$$

$$0.285 \text{ mol Na}_2SO_4 \cdot \frac{1 \text{ mol BaSO}_4}{1 \text{ mol Na}_2SO_4} = 0.285 \text{ mol BaSO}_4$$

$$0.285 \text{ mol BaSO}_4 \cdot \frac{233.4 \text{ g}}{\text{mol}} = 67 \text{ g}$$

28.

$H_2SO_4 + 2NaOH \rightarrow Na_2SO_4 + 2H_2O$

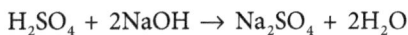

$655 \text{ mL} \cdot \dfrac{1 \text{ L}}{1000 \text{ mL}} = 0.655 \text{ L}$

$0.655 \text{ L} \cdot \dfrac{6.00 \text{ mol}}{\text{L}} = 3.930 \text{ mol NaOH}$

$3.930 \text{ mol NaOH} \cdot \dfrac{1 \text{ mol H}_2\text{SO}_4}{2 \text{ mol NaOH}} = 1.965 \text{ mol H}_2\text{SO}_4$

$1.965 \text{ mol H}_2\text{SO}_4 \cdot \dfrac{\text{L}}{12.0 \text{ mol}} = 0.164 \text{ L} \cdot \dfrac{1000 \text{ mL}}{1 \text{ L}} = 164 \text{ mL}$

29.

$Fe + 2HCl \rightarrow FeCl_2 + H_2$

$2.00 \times 10^3 \text{ L} \cdot \dfrac{2.0 \text{ mol}}{\text{L}} = 4.00 \times 10^3 \text{ mol HCl}$

$4.00 \times 10^3 \text{ mol HCl} \cdot \dfrac{1 \text{ mol H}_2}{2 \text{ mol HCl}} = 2.00 \times 10^3 \text{ mol H}_2$

$P = 100 \text{ kPa}$

$T = 273.15 \text{ K}$

$R = 8.314 \dfrac{\text{L} \cdot \text{kPa}}{\text{mol} \cdot \text{K}}$

$PV = nRT \rightarrow V = \dfrac{nRT}{P} = \dfrac{2.00 \times 10^3 \text{ mol} \cdot 8.314 \dfrac{\text{L} \cdot \text{kPa}}{\text{mol} \cdot \text{K}} \cdot 273.15 \text{ K}}{100 \text{ kPa}} = 45{,}000 \text{ L}$

---

30. a.

$10.0 \dfrac{\text{mol}}{\text{kg}} \cdot 1.00 \text{ kg} = 10.0 \text{ mol HCl}$

$10.0 \text{ mol HCl} \cdot \dfrac{36.46 \text{ g}}{\text{mol}} = 365 \text{ g}$

---

30. b.

$1.75 \text{ L} \cdot \dfrac{1000 \text{ mL}}{1 \text{ L}} \cdot 0.998 \dfrac{\text{g}}{\text{mL}} = 1747 \text{ g} \cdot \dfrac{1 \text{ kg}}{1000 \text{ g}} = 1.747 \text{ kg}$

$5.25 \dfrac{\text{mol}}{\text{kg}} \cdot 1.747 \text{ kg} = 9.172 \text{ mol NaOH}$

$9.172 \text{ mol NaOH} \cdot \dfrac{40.00 \text{ g}}{\text{mol}} = 367 \text{ g}$

31.

$$5.00\times10^2 \text{ mL}\cdot0.998 \ \frac{\text{g}}{\text{mL}} = 499.0 \text{ g}\cdot\frac{1 \text{ kg}}{1000 \text{ g}} = 0.4990 \text{ kg}$$

$$225 \text{ g}\cdot\frac{\text{mol}}{342.3 \text{ g}} = 0.6573 \text{ mol}$$

$$\frac{0.6573 \text{ mol } C_{12}H_{22}O_{11}}{0.4990 \text{ kg } H_2O} = 1.32 \ m$$

32.

$$235 \text{ g}\cdot\frac{\text{mol}}{164.1 \text{ g}} = 1.432 \text{ mol}$$

$$\frac{1.432 \text{ mol } Ca(NO_3)_2}{x \text{ kg } H_2O} = 2.0 \ \frac{\text{mol}}{\text{kg}}$$

$$x = \frac{1.432 \text{ mol } Ca(NO_3)_2}{2.0 \ \dfrac{\text{mol}}{\text{kg}}} = 0.7161 \text{ kg } H_2O\cdot\frac{1000 \text{ g}}{\text{kg}} = 716.1 \text{ g}$$

$$716.1 \text{ g}\cdot\frac{\text{mL}}{0.998 \text{ g}} = 718 \text{ mL}$$

Rounding this value to 2 sig digs gives 720 mL.

---

35. c.

$$MgCl_2 + Pb(NO_3)_2 \rightarrow Mg(NO_3)_2 + PbCl_2$$

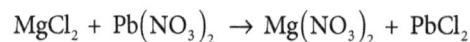

$$40.0 \text{ mL}\cdot\frac{1 \text{ L}}{1000 \text{ mL}}\cdot\frac{0.50 \text{ mol}}{\text{L}} = 0.0200 \text{ mol } MgCl_2$$

$$0.0200 \text{ mol } MgCl_2\cdot\frac{1 \text{ PbCl}_2}{1 \text{ mol } MgCl_2} = 0.0200 \text{ mol } PbCl_2$$

$$0.0200 \text{ mol } PbCl_2\cdot\frac{278.1 \text{ g}}{\text{mol}} = 5.6 \text{ g}$$

39.

$T_b = 99.974°C$

$K_b = 0.513 \dfrac{°C}{m}$

$\text{solute mass} = 0.4000 \text{ kg} \cdot \dfrac{1000 \text{ g}}{1 \text{ kg}} = 400.0 \text{ g } C_{12}H_{22}O_{11}$

$400.0 \text{ g } C_{12}H_{22}O_{11} \cdot \dfrac{\text{mol}}{342.30 \text{ g}} = 1.1686 \text{ mol}$

$1.100 \text{ L} \cdot \dfrac{1000 \text{ mL}}{1 \text{ L}} \cdot \dfrac{0.998 \text{ g}}{\text{mL}} \cdot \dfrac{1 \text{ kg}}{1000 \text{ g}} = 1.098 \text{ kg } H_2O$

$\text{molality} = \dfrac{1.1686 \text{ mol}}{1.098 \text{ kg}} = 1.0644 \ m$

$\Delta T_b = K_b m = 0.513 \dfrac{°C}{m} \cdot 1.0644 \ m = 0.546°C$

$T_b + \Delta T_b = 99.974°C + 0.546°C = 100.520°C$

40.

$T_b = 99.974°C$

$K_b = 0.513 \dfrac{°C}{m}$

First concentration:

$\Delta T_b = 101.75°C - 99.974°C = 1.776°C$

$\Delta T_b = K_b m \rightarrow m = \dfrac{\Delta T_b}{K_b} = \dfrac{1.776°C}{0.513 \dfrac{°C}{m}} = 3.46 \ m$

Second concentration:

$\Delta T_b = 100.99°C - 99.974°C = 1.016°C$

$\Delta T_b = K_b m \rightarrow m = \dfrac{\Delta T_b}{K_b} = \dfrac{1.016°C}{0.513 \dfrac{°C}{m}} = 1.98 \ m$

Third concentration:

$\Delta T_b = 103.17°C - 99.974°C = 3.196°C$

$\Delta T_b = K_b m \rightarrow m = \dfrac{\Delta T_b}{K_b} = \dfrac{3.196°C}{0.513 \dfrac{°C}{m}} = 6.23 \ m$

41.

$\Delta T_f = 18°C$

$K_f = 1.86 \dfrac{°C}{m}$

$\Delta T_f = K_f m \rightarrow m = \dfrac{\Delta T_f}{K_f} = \dfrac{18°C}{1.86 \dfrac{°C}{m}} = 9.7 \ m$

42.

$\Delta T_f = 2.93°C$

$K_f = 1.86 \dfrac{°C}{m}$

$\text{mass of solvent} = 775 \ g \cdot \dfrac{1 \ kg}{1000 \ g} = 0.775 \ kg$

$\Delta T_f = K_f m \rightarrow m = \dfrac{\Delta T_f}{K_f} = \dfrac{2.93°C}{1.86 \dfrac{°C}{m}} = 1.575 \ m$

$1.575 \ m = \dfrac{\text{moles } CO(NH_2)_2}{0.775 \ kg} \rightarrow \text{moles } CO(NH_2)_2 = 1.575 \ m \cdot 0.775 \ kg = 1.220 \ mol$

$1.220 \ mol \ CO(NH_2)_2 \cdot \dfrac{60.055 \ g}{mol} = 73.3 \ g$

43.

First, determine the mass of benzene:

s.g. $= 0.878$

$V = 2.50$ L

$\rho_{water} = 0.998 \; \dfrac{g}{mL}$

$0.878 \cdot 0.998 \; \dfrac{g}{mL} = 0.8762 \; \dfrac{g}{mL} \cdot \dfrac{1000 \; mL}{1 \; L} = 876.2 \; \dfrac{g}{L}$

$876.2 \; \dfrac{g}{L} \cdot 2.50 \; L = 2191 \; g \cdot \dfrac{1 \; kg}{1000 \; g} = 2.191 \; kg \; C_6H_6$

Next, determine the molality of the solution:

$\Delta T_b = 85.34°C - 80.1°C = 5.24°C$

$K_b = 2.64 \; \dfrac{°C}{m}$

$\Delta T_b = K_b m \rightarrow m = \dfrac{\Delta T_b}{K_b} = \dfrac{5.24°C}{2.64 \; \dfrac{°C}{m}} = 1.98 \; m$

Finally, put these together to determine the amount of CCl_4:

$1.98 \; m = \dfrac{moles \; CCl_4}{2.191 \; kg \; C_6H_6} \rightarrow moles \; CCl_4 = 1.98 \; m \cdot 2.191 \; kg = 4.338 \; mol$

$4.338 \; mol \; CCl_4 \cdot \dfrac{153.8 \; g}{mol} = 667 \; g$

This value must be rounded to 2 sig digs to give 670 g. This is because the calculation of ΔT_b using the addition rule gives 5.2°C (the 5.24°C has an extra digit), and this 2-digit value limits all subsequent results to 2 digits.

44.

Each mole of LiCl produces two moles of ions.

$$K_f = 1.86 \, \frac{°C}{m}$$

$$T_f = 0.00°C$$

$$575 \text{ g H}_2\text{O} \cdot \frac{1 \text{ kg}}{1000 \text{ g}} = 0.575 \text{ kg H}_2\text{O}$$

$$155 \text{ g} \cdot \frac{\text{mol}}{42.39 \text{ g}} = 3.656 \text{ mol LiCl}$$

$$3.656 \text{ mol LiCl} \cdot \frac{2 \text{ mol ions}}{1 \text{ mol LiCl}} = 7.312 \text{ mol ions}$$

$$m = \frac{7.312 \text{ mol ions}}{0.575 \text{ kg H}_2\text{O}} = 12.72 \, m$$

$$\Delta T_f = K_f m = 1.86 \, \frac{°C}{m} \cdot 12.72 \, m = 23.7°C$$

$$0.00°C - 23.7°C = -23.7°C$$

45.

$$K_f = 1.959 \, \frac{°C}{m}$$

$$T_f = -114.4°C$$

$$456 \text{ g C}_2\text{H}_5\text{OH} \cdot \frac{1 \text{ kg}}{1000 \text{ g}} = 0.456 \text{ kg C}_2\text{H}_5\text{OH}$$

$$75.0 \text{ g} \cdot \frac{\text{mol}}{74.55 \text{ g}} = 1.006 \text{ mol KCl}$$

$$1.006 \text{ mol KCl} \cdot \frac{2 \text{ mol ions}}{1 \text{ mol KCl}} = 2.012 \text{ mol ions}$$

$$m = \frac{2.012 \text{ mol ions}}{0.456 \text{ kg C}_2\text{H}_5\text{OH}} = 4.412 \, m$$

$$\Delta T_f = K_f m = 1.959 \, \frac{°C}{m} \cdot 4.412 \, m = 8.644°C$$

$$-114.4°C - 8.644°C = -123.0°C$$

46.

$$K_b = 3.22 \; \frac{°C}{m}$$

$$m = 0.040 \; m$$

$$0.040 \; m \; KI \cdot \frac{2 \; mol \; ions}{1 \; mol \; KI} = 0.080 \; m \; KI \; ions$$

$$\Delta T_b = K_b m = 3.22 \; \frac{°C}{m} \cdot 0.080 \; m = +0.26°C \; \text{(for KI)}$$

$$0.040 \; m \; MgCl_2 \cdot \frac{3 \; mol \; ions}{1 \; mol \; KI} = 0.120 \; m \; MgCl_2 \; ions$$

$$\Delta T_b = K_b m = 3.22 \; \frac{°C}{m} \cdot 0.120 \; m = +0.39°C \; \text{(for MgCl}_2\text{)}$$

$$0.040 \; m \; CaSO_4 \cdot \frac{2 \; mol \; ions}{1 \; mol \; KI} = 0.080 \; m \; CaSO_4 \; ions$$

$$\Delta T_b = K_b m = 3.22 \; \frac{°C}{m} \cdot 0.080 \; m = +0.26°C \; \text{(for CaSO}_4\text{; high ionic charges will likely}$$

make the actual change less than this prediction)

47.

$$T_f = 0.00°C$$

$$K_f = 1.86 \; \frac{°C}{m}$$

$$m = 0.020 \; m$$

For Na_3PO_4:

$$0.020 \; m \cdot \frac{4 \; mol \; ions}{1 \; mol \; KI} = 0.080 \; m \; ions$$

$$\Delta T_f = K_f m = 1.86 \; \frac{°C}{m} \cdot 0.080 \; m = 0.15°C$$

$$0.00 - 0.15°C = -0.15°C$$

For KCl:

$$0.020 \; m \cdot \frac{2 \; mol \; ions}{1 \; mol \; KI} = 0.040 \; m \; ions$$

$$\Delta T_f = K_f m = 1.86 \; \frac{°C}{m} \cdot 0.040 \; m = 0.07°C$$

$$0.00 - 0.07°C = -0.07°C$$

For $C_6H_{12}O_6$:

$$0.020 \; m \cdot \frac{1 \; mol \; ions}{1 \; mol \; KI} = 0.020 \; m \; ions$$

$$\Delta T_f = K_f m = 1.86 \; \frac{°C}{m} \cdot 0.020 \; m = 0.04°C$$

$$0.00 - 0.04°C = -0.04°C$$

For $CaCl_2$:

$$0.020 \; m \cdot \frac{3 \; mol \; ions}{1 \; mol \; KI} = 0.060 \; m \; ions$$

$$\Delta T_f = K_f m = 1.86 \; \frac{°C}{m} \cdot 0.060 \; m = 0.11°C$$

$$0.00 - 0.11°C = -0.11°C$$

48.

$m = 2.00 \ m$

$K_f = 1.86 \ \dfrac{°C}{m}$

$2.00 \ m \cdot \dfrac{2 \ \text{mol ions}}{1 \ \text{mol HNO}_3} = 4.00 \ m \ \text{ions}$

$\Delta T_f = K_f m = 1.86 \ \dfrac{°C}{m} \cdot 4.00 \ m = -7.44°C$

49.

$m = 1.30 \ \text{g}$

$V = 2.24 \ \text{L}$

$T = 6.85°C \rightarrow 6.85°C + 273.15 = 280.00 \ \text{K}$

$P = 1.14 \ \text{atm}$

$R = 0.08206 \ \dfrac{\text{L} \cdot \text{atm}}{\text{mol} \cdot \text{K}}$

$PV = nRT \rightarrow n = \dfrac{PV}{RT} = \dfrac{1.14 \ \text{atm} \cdot 2.24 \ \text{L}}{0.08206 \ \dfrac{\text{L} \cdot \text{atm}}{\text{mol} \cdot \text{K}} \cdot 280.00 \ \text{K}} = 0.1111 \ \text{mol}$

$M = \dfrac{1.30 \ \text{g}}{0.1111 \ \text{mol}} = 11.7 \ \dfrac{\text{g}}{\text{mol}}$

50.

$CH_4(g) + 2O_2(g) \rightarrow CO_2(g) + 2H_2O(g)$

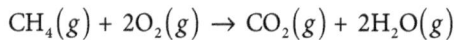

$17.5 \ \text{L CH}_4 \cdot \dfrac{1 \ \text{mol CO}_2}{1 \ \text{mol CH}_4} = 17.5 \ \text{L CO}_2$

$17.5 \ \text{L CH}_4 \cdot \dfrac{2 \ \text{mol H}_2\text{O}}{1 \ \text{mol CH}_4} = 35.0 \ \text{L H}_2\text{O}$

52.

Using F_w to indicate the weight,

$F_w = 12,500 \ \text{lb} \cdot \dfrac{4.448 \ \text{N}}{1 \ \text{lb}} = 55,600 \ \text{N}$

$l = 10.0 \ \text{in} \cdot \dfrac{2.54 \ \text{cm}}{\text{in}} \cdot \dfrac{1 \ \text{m}}{100 \ \text{cm}} = 0.254 \ \text{m}$

$w = 10.0 \ \text{in} \cdot \dfrac{2.54 \ \text{cm}}{\text{in}} \cdot \dfrac{1 \ \text{m}}{100 \ \text{cm}} = 0.254 \ \text{m}$

$A = l \cdot w = 0.254 \ \text{m} \cdot 0.254 \ \text{m} = 0.06452 \ \text{m}^2$

$P = \dfrac{F}{A} = \dfrac{55,600 \ \text{N}}{0.06452 \ \text{m}^2} = 861,750 \ \text{Pa} \cdot \dfrac{1 \ \text{atm}}{101,325 \ \text{Pa}} = 8.50 \ \text{atm}$

53.

$$100.0 \text{ kg} \cdot \frac{1000 \text{ g}}{1 \text{ kg}} \cdot \frac{\text{mol}}{12.011 \text{ g}} = 8325.7 \text{ mol C}$$

$$8325.7 \text{ mol C} \cdot \frac{1 \text{ mol CH}_4}{2 \text{ mol C}} = 4162.9 \text{ mol CH}_4$$

$$4162.9 \text{ mol CH}_4 \cdot \frac{16.043 \text{ g}}{\text{mol}} = 66,780 \text{ g}$$

$$66,780 \text{ g} \cdot \frac{1 \text{ kg}}{1000 \text{ g}} \cdot 0.82 = 55 \text{ kg}$$

54.

$$125.0 \text{ mol C}_7\text{H}_6\text{O}_3 \cdot \frac{1 \text{ mol C}_9\text{H}_8\text{O}_4}{1 \text{ mol C}_7\text{H}_6\text{O}_3} = 125.0 \text{ mol C}_9\text{H}_8\text{O}_4$$

$$125.0 \text{ mol C}_9\text{H}_8\text{O}_4 \cdot \frac{180.16 \text{ g}}{\text{mol}} = 22,520.0 \text{ g}$$

$$22,520.0 \text{ g} \cdot \frac{1 \text{ kg}}{1000 \text{ g}} = 22.52 \text{ kg}$$

55.

$$\frac{7 \cdot 12.011}{138.123} = 0.60871 \rightarrow 60.871\% \text{ C}$$

$$\frac{6 \cdot 1.0079}{138.123} = 0.043783 \rightarrow 4.3783\% \text{ H}$$

$$\frac{3 \cdot 15.9994}{138.123} = 0.347503 \rightarrow 34.7503\% \text{ O}$$

Chapter 9

Note: In later printings, question 23 was renumbered to be question 26 to be included in Section 9.3. Accordingly, items below formerly numbeed as 24–26 have been renumbered 23–25.

23. a.

$$2HNO_3 + Na_2CO_3 \rightarrow CO_2 + 2NaNO_3 + H_2O$$

$$196 \text{ g Na}_2CO_3 \cdot \frac{\text{mol}}{105.99 \text{ g}} = 1.849 \text{ mol Na}_2CO_3$$

$$1.849 \text{ mol Na}_2CO_3 \cdot \frac{1 \text{ mol CO}_2}{1 \text{ mol Na}_2CO_3} = 1.849 \text{ mol CO}_2$$

$$1.849 \text{ mol CO}_2 \cdot \frac{44.01 \text{ g}}{\text{mol}} = 81.4 \text{ g CO}_2$$

$$P = 100 \text{ kPa}$$

$$T = 273.15 \text{ K}$$

$$R = 8.314 \frac{\text{L} \cdot \text{kPa}}{\text{mol} \cdot \text{K}}$$

$$PV = nRT \rightarrow V = \frac{nRT}{P} = \frac{1.849 \text{ mol} \cdot 8.314 \dfrac{\text{L} \cdot \text{kPa}}{\text{mol} \cdot \text{K}} \cdot 273.15 \text{ K}}{100 \text{ kPa}} = 42.0 \text{ L CO}_2$$

---

23. b.

$$1.849 \text{ mol Na}_2CO_3 \cdot \frac{2 \text{ mol HNO}_3}{1 \text{ mol Na}_2CO_3} = 3.698 \text{ mol HNO}_3$$

$$3.698 \text{ mol HNO}_3 \cdot \frac{\text{L}}{6.00 \text{ mol}} = 0.616 \text{ L} \cdot \frac{1000 \text{ mL}}{1 \text{ L}} = 616 \text{ mL}$$

---

24. a.

$$Cu + H_2SO_4 \rightarrow CuSO_4 + H_2$$

$$125 \text{ mL} \cdot \frac{1 \text{ L}}{1000 \text{ mL}} = 0.125 \text{ L}$$

$$0.125 \text{ L} \cdot \frac{8.0 \text{ mol}}{\text{L}} = 1.000 \text{ mol H}_2SO_4$$

$$1.000 \text{ mol H}_2SO_4 \cdot \frac{1 \text{ mol CuSO}_4}{1 \text{ mol H}_2SO_4} = 1.000 \text{ mol CuSO}_4$$

$$1.000 \text{ mol CuSO}_4 \cdot \frac{159.61 \text{ g}}{\text{mol}} = 160 \text{ g}$$

---

24. b.

$$1.000 \text{ mol } H_2SO_4 \cdot \frac{1 \text{ mol } H_2}{1 \text{ mol } H_2SO_4} = 1.000 \text{ mol } H_2$$

$P = 100 \text{ kPa}$

$T = 273.15 \text{ K}$

$R = 8.314 \dfrac{L \cdot kPa}{mol \cdot K}$

$$PV = nRT \rightarrow V = \frac{nRT}{P} = \frac{1.000 \text{ mol} \cdot 8.314 \dfrac{L \cdot kPa}{mol \cdot K} \cdot 273.15 \text{ K}}{100 \text{ kPa}} = 22.7 \text{ L}$$

---

25. a.

$$CaCO_3 + 2HCl \rightarrow CaCl_2 + CO_2 + H_2O$$

$$V = 2.5 \times 10^4 \text{ mL} \cdot \frac{1 \text{ L}}{1000 \text{ mL}} = 25 \text{ L } CO_2$$

$P = 100 \text{ kPa}$

$T = 273.15$

$R = 8.314 \dfrac{L \cdot kPa}{mol \cdot K}$

$$PV = nRT \rightarrow n = \frac{PV}{RT} = \frac{100 \text{ kPa} \cdot 25 \text{ L}}{8.314 \dfrac{L \cdot kPa}{mol \cdot K} \cdot 273.15} = 1.10 \text{ mol } CO_2$$

$$1.10 \text{ mol } CO_2 \cdot \frac{1 \text{ mol } CaCO_3}{1 \text{ mol } CO_2} = 1.10 \text{ mol } CaCO_3$$

$$1.10 \text{ mol } CaCO_3 \cdot \frac{100.1 \text{ g}}{mol} = 110 \text{ g}$$

---

25. b

$$1.10 \text{ mol } CO_2 \cdot \frac{2 \text{ mol } HCl}{1 \text{ mol } CO_2} = 2.20 \text{ mol } HCl$$

$$2.20 \text{ mol } HCl \cdot \frac{L}{2.00 \text{ mol}} = 1.1 \text{ L}$$

---

32. a

$$[H_3O^+] = 1.56 \times 10^{-3} \ M$$

$$[H_3O^+] \cdot [OH^-] = 1.0 \times 10^{-14} \ M^2$$

$$[OH^-] = \frac{1.0 \times 10^{-14} \ M^2}{[H_3O^+]} = \frac{1.0 \times 10^{-14} \ M^2}{1.56 \times 10^{-3} \ M} = 6.4 \times 10^{-12} \ M$$

---

32. b.

Each mole of $Sr(OH)_2$ contributes two moles of OH^- ions.

$$\left[OH^-\right] = 2 \cdot 0.0110 \ M = 0.0220 \ M$$

$$\left[H_3O^+\right] \cdot \left[OH^-\right] = 1.0 \times 10^{-14} \ M^2$$

$$\left[H_3O^+\right] = \frac{1.0 \times 10^{-14} \ M^2}{\left[OH^-\right]} = \frac{1.0 \times 10^{-14} \ M^2}{0.0220 \ M} = 4.5 \times 10^{-13} \ M$$

32. c.

$$\left[OH^-\right] = 2.50 \ M$$

$$\left[H_3O^+\right] \cdot \left[OH^-\right] = 1.0 \times 10^{-14} \ M^2$$

$$\left[H_3O^+\right] = \frac{1.0 \times 10^{-14} \ M^2}{\left[OH^-\right]} = \frac{1.0 \times 10^{-14} \ M^2}{2.50 \ M} = 4.0 \times 10^{-15} \ M$$

32. d.

$$\left[H_3O^+\right] = 8.911 \times 10^{-5} \ M$$

$$\left[H_3O^+\right] \cdot \left[OH^-\right] = 1.0 \times 10^{-14} \ M^2$$

$$\left[OH^-\right] = \frac{1.0 \times 10^{-14} \ M^2}{\left[H_3O^+\right]} = \frac{1.0 \times 10^{-14} \ M^2}{8.911 \times 10^{-5} \ M} = 1.1 \times 10^{-10} \ M$$

33.

In each case, the number of sig digs in the result is determined by two principles: 1) the number of sig digs in a number you are taking a logarithm of must equal the number of digits to the right of the decimal in your logarithm; 2) after taking a logarithm to obtain pH or pOH, the other one is obtained from an equation using either the addition rule and the sig digs are determined accordingly.

33. a.

$$pOH = -\log\left[OH^-\right] = -\log 0.0250 = 1.602$$

$$pH + pOH = 14.0$$

$$pH = 14.0 - pOH = 14.0 - 1.602 = 12.4$$

33. b.

$$pH = -\log\left[H_3O^+\right] = -\log 6.0 = -0.78$$

$$pH + pOH = 14.0$$

$$pOH = 14.0 - pH = 14.0 - (-0.78) = 14.8$$

33. c.

$$pH = -\log[H_3O^+] = -\log 12 = -1.08$$
$$pH + pOH = 14.0$$
$$pOH = 14.0 - pH = 14.0 - (-1.08) = 15.1$$

---

33. d.

$$[OH^-] = 2 \cdot 2.03 \times 10^{-2} \ M = 4.06 \times 10^{-2} \ M$$
$$pOH = -\log[OH^-] = -\log(4.06 \times 10^{-2}) = 1.391$$
$$pH + pOH = 14.0$$
$$pH = 14.0 - pOH = 14.0 - (1.391) = 12.6$$

34.

When computing with antilogs, the number of sig digs in the antilog must equal the number of digits to the right of the decimal in the value you started with. After using an antilog obtain $[H_3O^+]$ or $[OH^-]$, the other concentration is obtained from an equation using the multiplication rule and the sig digs are determined accordingly.

34. a.

$$pH = 1.00$$
$$[H_3O^+] = 10^{-pH} = 10^{-1.00} = 0.10 \ M$$
$$[H_3O^+] \cdot [OH^-] = 1.0 \times 10^{-14} \ M^2$$
$$[OH^-] = \frac{1.0 \times 10^{-14} \ M^2}{[H_3O^+]} = \frac{1.0 \times 10^{-14} \ M^2}{0.10 \ M} = 1.0 \times 10^{-13} \ M$$

---

34. b.

$$pH = 3.5$$
$$[H_3O^+] = 10^{-pH} = 10^{-3.5} = 0.0003 \ M$$
$$[H_3O^+] \cdot [OH^-] = 1.0 \times 10^{-14} \ M^2$$
$$[OH^-] = \frac{1.0 \times 10^{-14} \ M^2}{[H_3O^+]} = \frac{1.0 \times 10^{-14} \ M^2}{0.0003 \ M} = 3 \times 10^{-11} \ M$$

---

34. c.

$$pOH = 10.0$$
$$[OH^-] = 10^{-pOH} = 10^{-10.0} = 1 \times 10^{-10} \ M$$
$$[H_3O^+] \cdot [OH^-] = 1.0 \times 10^{-14} \ M^2$$
$$[H_3O^+] = \frac{1.0 \times 10^{-14} \ M^2}{[OH^-]} = \frac{1.0 \times 10^{-14} \ M^2}{1 \times 10^{-10} \ M} = 1 \times 10^{-4} \ M$$

---

34. d.

$pH = 11.88$

$\left[H_3O^+ \right] = 10^{-pH} = 10^{-11.88} = 1.3 \times 10^{-12} \ M$

$\left[H_3O^+ \right] \cdot \left[OH^- \right] = 1.0 \times 10^{-14} \ M^2$

$\left[OH^- \right] = \dfrac{1.0 \times 10^{-14} \ M^2}{\left[H_3O^+ \right]} = \dfrac{1.0 \times 10^{-14} \ M^2}{1.3 \times 10^{-12} \ M} = 0.0077 \ M$

---

34. e.

$pOH = 2.1$

$\left[OH^- \right] = 10^{-pOH} = 10^{-2.1} = 0.008 \ M$

$\left[H_3O^+ \right] \cdot \left[OH^- \right] = 1.0 \times 10^{-14} \ M^2$

$\left[H_3O^+ \right] = \dfrac{1.0 \times 10^{-14} \ M^2}{\left[OH^- \right]} = \dfrac{1.0 \times 10^{-14} \ M^2}{0.008 \ M} = 1 \times 10^{-12} \ M$

---

34. f.

$pOH = 7.0$

$\left[OH^- \right] = 10^{-pOH} = 10^{-7.0} = 1 \times 10^{-7} \ M$

$\left[H_3O^+ \right] \cdot \left[OH^- \right] = 1.0 \times 10^{-14} \ M^2$

$\left[H_3O^+ \right] = \dfrac{1.0 \times 10^{-14} \ M^2}{\left[OH^- \right]} = \dfrac{1.0 \times 10^{-14} \ M^2}{1 \times 10^{-7} \ M} = 1 \times 10^{-7} \ M$

---

34. g.

$pOH = 4.00$

$\left[OH^- \right] = 10^{-pOH} = 10^{-4.00} = 1.0 \times 10^{-4} \ M$

$\left[H_3O^+ \right] \cdot \left[OH^- \right] = 1.0 \times 10^{-14} \ M^2$

$\left[H_3O^+ \right] = \dfrac{1.0 \times 10^{-14} \ M^2}{\left[OH^- \right]} = \dfrac{1.0 \times 10^{-14} \ M^2}{1.0 \times 10^{-4} \ M} = 1.0 \times 10^{-10} \ M$

---

34. h.

$pH = 14.0$

$\left[H_3O^+ \right] = 10^{-pH} = 10^{-14.0} = 1 \times 10^{-14} \ M$

$\left[H_3O^+ \right] \cdot \left[OH^- \right] = 1.0 \times 10^{-14} \ M^2$

$\left[OH^- \right] = \dfrac{1.0 \times 10^{-14} \ M^2}{\left[H_3O^+ \right]} = \dfrac{1.0 \times 10^{-14} \ M^2}{1 \times 10^{-14} \ M} = 1 \ M$

---

35. a.

$pH = 5.6$

$pH + pOH = 14.0$

$pOH = 14.0 - pH = 14.0 - 5.6 = 8.4$

$\left[H_3O^+\right] = 10^{-pH} = 10^{-5.6} = 3 \times 10^{-6} \ M$

$\left[OH^-\right] = 10^{-pOH} = 10^{-8.4} = 4 \times 10^{-9} \ M$

Alternately, this calculation for $\left[OH^-\right]$ leads to a slightly different result due to limits on precision:

$\left[H_3O^+\right] \cdot \left[OH^-\right] = 1.0 \times 10^{-14} \ M^2$

$\left[OH^-\right] = \dfrac{1.0 \times 10^{-14} \ M^2}{\left[H_3O^+\right]} = \dfrac{1.0 \times 10^{-14} \ M^2}{3 \times 10^{-6} \ M} = 3 \times 10^{-9} \ M$

---

35. b.

$\left[OH^-\right] = 5.5 \times 10^{-10} \ M$

$\left[H_3O^+\right] \cdot \left[OH^-\right] = 1.0 \times 10^{-14} \ M^2$

$\left[H_3O^+\right] = \dfrac{1.0 \times 10^{-14} \ M^2}{\left[OH^-\right]} = \dfrac{1.0 \times 10^{-14} \ M^2}{5.5 \times 10^{-10} \ M} = 1.8 \times 10^{-5} \ M$

$pOH = -\log\left[OH^-\right] = -\log 5.5 \times 10^{-10} = 9.26$

$pH + pOH = 14.0$

$pH = 14.0 - pOH = 14.0 - 9.26 = 4.7$

---

35. c.

$\left[H_3O^+\right] = 1.0 \times 10^{-9} \ M$

$\left[H_3O^+\right] \cdot \left[OH^-\right] = 1.0 \times 10^{-14} \ M^2$

$\left[OH^-\right] = \dfrac{1.0 \times 10^{-14} \ M^2}{\left[H_3O^+\right]} = \dfrac{1.0 \times 10^{-14} \ M^2}{1.0 \times 10^{-9} \ M} = 1.0 \times 10^{-5} \ M$

$pH = -\log\left[H_3O^+\right] = -\log\left(1.0 \times 10^{-9}\right) = 9.00$

For highest precision, calculate pOH this way:

$pOH = -\log\left[OH^-\right] = -\log\left(1.0 \times 10^{-5}\right) = 5.00$

---

35. d.

$pOH = 11.5$

$pH + pOH = 14.0$

$pH = 14.0 - pOH = 14.0 - 11.5 = 2.5$

$\left[OH^- \right] = 10^{-pOH} = 10^{-11.5} = 3 \times 10^{-12} \ M$

$\left[H_3O^+ \right] = 10^{-pH} = 10^{-2.5} = 0.003 \ M$

Alternately:

$\left[H_3O^+ \right] \cdot \left[OH^- \right] = 1.0 \times 10^{-14} \ M^2$

$\left[H_3O^+ \right] = \dfrac{1.0 \times 10^{-14} \ M^2}{\left[OH^- \right]} = \dfrac{1.0 \times 10^{-14} \ M^2}{3 \times 10^{-12} \ M} = 0.003 \ M$

---

35. e.

$\left[H_3O^+ \right] = 4.5 \times 10^{-3} \ M$

$\left[H_3O^+ \right] \cdot \left[OH^- \right] = 1.0 \times 10^{-14} \ M^2$

$\left[OH^- \right] = \dfrac{1.0 \times 10^{-14} \ M^2}{\left[H_3O^+ \right]} = \dfrac{1.0 \times 10^{-14} \ M^2}{4.5 \times 10^{-3} \ M} = 2.2 \times 10^{-12} \ M$

$pH = -\log \left[H_3O^+ \right] = -\log \left(4.5 \times 10^{-3} \right) = 2.35$

For highest precision, calculate pOH this way:

$pOH = -\log \left[OH^- \right] = -\log \left(2.2 \times 10^{-12} \right) = 11.66$

---

35. f.

$pOH = 1.00$

$pH + pOH = 14.0$

$pH = 14.0 - pOH = 14.0 - 1.00 = 13.0$

$\left[OH^- \right] = 10^{-pOH} = 10^{-1.00} = 0.10 \ M$

For highest precision, calculate $\left[H_3O^+ \right]$ this way:

$\left[H_3O^+ \right] \cdot \left[OH^- \right] = 1.0 \times 10^{-14} \ M^2$

$\left[H_3O^+ \right] = \dfrac{1.0 \times 10^{-14} \ M^2}{\left[OH^- \right]} = \dfrac{1.0 \times 10^{-14} \ M^2}{0.10 \ M} = 1.0 \times 10^{-13} \ M$

---

35. g.

$pOH = 7.22$

$pH + pOH = 14.0$

$pH = 14.0 - pOH = 14.0 - 7.22 = 6.8$

$\left[OH^-\right] = 10^{-pOH} = 10^{-7.22} = 6.0 \times 10^{-8} \ M$

$\left[H_3O^+\right] = 10^{-pH} = 10^{-6.8} = 2 \times 10^{-7} \ M$

Alternately, this calculation for $\left[H_3O^+\right]$ allows an extra sig dig:

$\left[H_3O^+\right] \cdot \left[OH^-\right] = 1.0 \times 10^{-14} \ M^2$

$\left[H_3O^+\right] = \dfrac{1.0 \times 10^{-14} \ M^2}{\left[OH^-\right]} = \dfrac{1.0 \times 10^{-14} \ M^2}{6.0 \times 10^{-8} \ M} = 1.7 \times 10^{-7} \ M$

---

35. h.

$pH = 13.50$

$pH + pOH = 14.0$

$pOH = 14.0 - pH = 14.0 - 13.50 = 0.5$

$\left[H_3O^+\right] = 10^{-pH} = 10^{-13.50} = 3.2 \times 10^{-14} \ M$

$\left[OH^-\right] = 10^{-pOH} = 10^{-0.5} = 0.3 \ M$

Alternately, this calculation for $\left[OH^-\right]$ allows an extra digit of precision:

$\left[H_3O^+\right] \cdot \left[OH^-\right] = 1.0 \times 10^{-14} \ M^2$

$\left[OH^-\right] = \dfrac{1.0 \times 10^{-14} \ M^2}{\left[H_3O^+\right]} = \dfrac{1.0 \times 10^{-14} \ M^2}{3.2 \times 10^{-14} \ M} = 0.31 \ M$

---

35. i.

$pH = 2.0$

$pH + pOH = 14.0$

$pOH = 14.0 - pH = 14.0 - 2.0 = 12.0$

$\left[H_3O^+\right] = 10^{-pH} = 10^{-2.0} = 0.01 \ M$

$\left[OH^-\right] = 10^{-pOH} = 10^{-12.0} = 1 \times 10^{-12} \ M$

Alternately:

$\left[H_3O^+\right] \cdot \left[OH^-\right] = 1.0 \times 10^{-14} \ M^2$

$\left[OH^-\right] = \dfrac{1.0 \times 10^{-14} \ M^2}{\left[H_3O^+\right]} = \dfrac{1.0 \times 10^{-14} \ M^2}{0.01 \ M} = 1 \times 10^{-12} \ M$

---

35. j.

$$\left[OH^-\right]=3.2\times10^{-2}\ M$$

$$\left[H_3O^+\right]\cdot\left[OH^-\right]=1.0\times10^{-14}\ M^2$$

$$\left[H_3O^+\right]=\frac{1.0\times10^{-14}\ M^2}{\left[OH^-\right]}=\frac{1.0\times10^{-14}\ M^2}{3.2\times10^{-2}\ M}=3.1\times10^{-13}\ M$$

$$pOH=-\log\left[OH^-\right]=-\log\left(3.2\times10^{-2}\ M\right)=1.49$$

$$pH=-\log\left[H_3O^+\right]=-\log\left(3.1\times10^{-13}\ M\right)=12.51$$

---

36. a.

$$24.05\ mL\cdot\frac{1\ L}{1000\ mL}\cdot1.65\times10^{-3}\ M=3.968\times10^{-5}\ mol\ NaOH$$

$$3.968\times10^{-5}\ mol\ NaOH\rightarrow3.968\times10^{-5}\ mol\ HNO_3$$

$$M_{HNO_3}\cdot18.55\ mL\cdot\frac{1\ L}{1000\ mL}=3.968\times10^{-5}\ mol\ HNO_3$$

$$M_{HNO_3}=\frac{3.968\times10^{-5}\ mol\ HNO_3}{0.01855\ L}=2.14\times10^{-3}\ M\ HNO_3$$

---

36. b.

$$21.00\ mL\cdot\frac{1\ L}{1000\ mL}\cdot1.500\ M=0.03150\ mol\ HClO_3$$

$$0.03150\ mol\ HClO_3\rightarrow0.03150\ mol\ KOH$$

$$M_{KOH}\cdot23.78\ mL\cdot\frac{1\ L}{1000\ mL}=0.03150\ mol\ KOH$$

$$M_{KOH}=\frac{0.03150\ mol\ KOH}{0.02378\ L}=1.325\ M\ KOH$$

---

36. c.

$$20.87\ mL\cdot\frac{1\ L}{1000\ mL}\cdot4.077\times10^{-2}\ M=8.509\times10^{-4}\ mol\ HClO_4$$

$Ca(OH)_2$ provides two moles of OH^- ions for each mole of $Ca(OH)_2$. So only half as many moles of $Ca(OH)_2$ are required to neutralize $HClO_4$.

$$\frac{8.509\times10^{-4}\ mol\ HClO_4}{2}\rightarrow4.254\times10^{-4}\ mol\ Ca(OH)_2$$

$$M_{Ca(OH)_2}\cdot22.94\ mL\cdot\frac{1\ L}{1000\ mL}=4.254\times10^{-4}\ mol\ Ca(OH)_2$$

$$M_{Ca(OH)_2}=\frac{4.254\times10^{-4}\ mol\ Ca(OH)_2}{0.02294\ L}=0.01855\ M\ Ca(OH)_2$$

---

36. d.

$$20.05 \text{ mL}\cdot\frac{1\text{ L}}{1000\text{ mL}}\cdot 3.050\ M = 0.06115 \text{ mol LiOH}$$

$$0.06115 \text{ mol LiOH} \rightarrow 0.06115 \text{ mol HCl}$$

$$M_{\text{HCl}}\cdot 22.00 \text{ mL}\cdot\frac{1\text{ L}}{1000\text{ mL}} = 0.06115 \text{ mol HCl}$$

$$M_{\text{HCl}} = \frac{0.06115 \text{ mol HCl}}{0.02200 \text{ L}} = 2.780\ M \text{ HCl}$$

37.

$$23.09 \text{ mL}\cdot\frac{1\text{ L}}{1000\text{ mL}}\cdot 5.06\times10^{-3}\ M = 1.1684\times10^{-4} \text{ mol HCl}$$

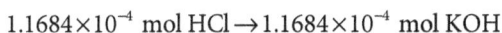

$$1.1684\times10^{-4} \text{ mol HCl} \rightarrow 1.1684\times10^{-4} \text{ mol KOH}$$

$$M_{\text{KOH}}\cdot 20.07 \text{ mL}\cdot\frac{1\text{ L}}{1000\text{ mL}} = 1.1684\times10^{-4} \text{ mol KOH}$$

$$M_{\text{KOH}} = \frac{1.1684\times10^{-4} \text{ mol KOH}}{0.02007 \text{ L}} = 5.82\times10^{-3}\ M \text{ KOH}$$

38.

$$22.01 \text{ mL}\cdot\frac{1\text{ L}}{1000\text{ mL}}\cdot 0.0231\ M = 5.084\times10^{-4} \text{ mol Ba(OH)}_2$$

Ba(OH)_2 provides two moles of OH^- ions for each mole of Ba(OH)_2. So twice as many moles of HCl will be neutralized.

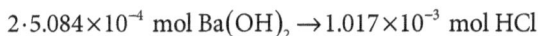

$$2\cdot 5.084\times10^{-4} \text{ mol Ba(OH)}_2 \rightarrow 1.017\times10^{-3} \text{ mol HCl}$$

$$M_{\text{HCl}}\cdot 27.0 \text{ mL}\cdot\frac{1\text{ L}}{1000\text{ mL}} = 1.017\times10^{-3} \text{ mol HCl}$$

$$M_{\text{HCl}} = \frac{1.017\times10^{-3} \text{ mol HCl}}{0.0270 \text{ L}} = 0.0377\ M \text{ HCl}$$

39.

$$23.90 \text{ mL}\cdot\frac{1\text{ L}}{1000\text{ mL}}\cdot 2.5\times10^{-3}\ M = 5.98\times10^{-5} \text{ mol NaOH}$$

5.98×10^{-5} mol NaOH neutralizes 5.98×10^{-5} mol acid, so 5.98×10^{-5} mol acid ionized.

$$20.00 \text{ mL}\cdot\frac{1\text{ L}}{1000\text{ mL}}\cdot 0.04098\ M = 8.196\times10^{-4} \text{ mol acid present}$$

$$\frac{5.98\times10^{-5} \text{ mol acid ionized}}{8.196\times10^{-4} \text{ mol acid present}} = 0.073 \rightarrow 7.3\% \text{ ionization}$$

40.

$$\frac{1.77 \text{ g}}{100 \text{ mL}}\cdot\frac{1000 \text{ mL}}{1\text{ L}}\cdot\frac{\text{mol}}{121.64 \text{ g}} = 0.146\ M \text{ (maximum concentration)}$$

41.

$$38.95 \text{ g} \cdot \frac{\text{mol}}{342.299 \text{ g}} = 0.11379 \text{ mol}$$

$$\frac{0.11379 \text{ mol}}{250.0 \text{ mL}} \cdot \frac{1000 \text{ mL}}{1 \text{ L}} = 0.4552 \text{ M}$$

42.

$$n = 37.05 \text{ g} \cdot \frac{\text{mol}}{18.015 \text{ g}} = 2.0566 \text{ mol}$$

$$C_{ice} = 0.0364 \ \frac{\text{kJ}}{\text{mol} \cdot \text{K}}$$

$$C_{water} = 0.0752 \ \frac{\text{kJ}}{\text{mol} \cdot \text{K}}$$

$$H_f = 6.01 \ \frac{\text{kJ}}{\text{mol}}$$

$$\Delta T_{ice} = 0.0°C - (-18.0°C) = 18.0°C = 18.0 \text{ K}$$

$$\Delta T_{water} = 20.0°C - 0.0°C = 20.0°C = 20.0 \text{ K}$$

To make the sig digs clear for the final addition step, we will list the results of the heat calculations without any extra sig digs.

To warm the ice to 0°C:

$$Q = Cn\Delta T = 0.0364 \ \frac{\text{kJ}}{\text{mol} \cdot \text{K}} \cdot 2.0566 \text{ mol} \cdot 18.0 \text{ K} = 1.35 \text{ kJ}$$

To melt the ice:

$$Q = nH_f = 2.0566 \text{ mol} \cdot 6.01 \ \frac{\text{kJ}}{\text{mol}} = 12.4 \text{ kJ}$$

To warm the water:

$$Q = Cn\Delta T = 0.0752 \ \frac{\text{kJ}}{\text{mol} \cdot \text{K}} \cdot 2.0566 \text{ mol} \cdot 20.0 \text{ K} = 3.09 \text{ kJ}$$

$$Q_{total} = 1.35 \text{ kJ} + 12.4 \text{ kJ} + 3.09 \text{ kJ} = 16.8 \text{ kJ}$$

---

43. a.

$$NaOH + HNO_3 \rightarrow NaNO_3 + H_2O$$

$$35.55 \text{ g NaOH} \cdot \frac{\text{mol}}{39.997 \text{ g}} = 0.88881 \text{ mol NaOH}$$

$$157.0 \text{ mL HNO}_3 \cdot \frac{1 \text{ L}}{1000 \text{ mL}} \cdot \frac{4.777 \text{ mol}}{\text{L}} = 0.74999 \text{ mol HNO}_3$$

The mole ratio is 1:1, so HNO_3 is the limiting reactant.

---

43. b.

$$0.74999 \text{ mol HNO}_3 \cdot \frac{1 \text{ mol NaNO}_3}{1 \text{ mol HNO}_3} = 0.7500 \text{ mol NaNO}_3$$

---

43. c.

$$0.74999 \text{ mol HNO}_3 \cdot \frac{1 \text{ mol H}_2\text{O}}{1 \text{ mol HNO}_3} = 0.7500 \text{ mol H}_2\text{O}$$

43. d.

$$0.8888 \text{ mol NaOH} - 0.7500 \text{ mol NaOH} = 0.1388 \text{ mol}$$

45. c.

$$25.50 \text{ mL} \cdot \frac{1 \text{ L}}{1000 \text{ mL}} \cdot \frac{0.050 \text{ mol}}{\text{L}} = 0.00128 \text{ mol Na}_2\text{CO}_3$$

$$0.00128 \text{ mol Na}_2\text{CO}_3 \cdot \frac{1 \text{ mol BaCO}_3}{1 \text{ mol Na}_2\text{CO}_3} = 0.00128 \text{ mol BaCO}_3$$

$$0.00128 \text{ mol BaCO}_3 \cdot \frac{197.336 \text{ g}}{\text{mol}} = 0.25 \text{ g}$$

46.

$$M_{CO_2} = 44.0 \ \frac{\text{g}}{\text{mol}}$$

$$M_{HCl} = 36.5 \ \frac{\text{g}}{\text{mol}}$$

$$CO_2 \text{ effusion rate} = 2.1 \times 10^{-3} \ \frac{\text{mol}}{\text{s}}$$

$$\frac{CO_2 \text{ effusion rate}}{HCl \text{ effusion rate}} = \frac{\sqrt{M_{HCl}}}{\sqrt{M_{CO_2}}}$$

$$HCl \text{ effusion rate} = CO_2 \text{ effusion rate} \cdot \frac{\sqrt{M_{CO_2}}}{\sqrt{M_{HCl}}} = 2.1 \times 10^{-3} \ \frac{\text{mol}}{\text{s}} \cdot \frac{\sqrt{44.0 \ \frac{\text{g}}{\text{mol}}}}{\sqrt{36.5 \ \frac{\text{g}}{\text{mol}}}} = 0.0023 \ \frac{\text{mol}}{\text{s}}$$

47.

$Cu + H_2SO_4 \rightarrow CuSO_4 + H_2$

$V = 109 \text{ mL H}_2 \cdot \dfrac{1 \text{ L}}{1000 \text{ mL}} = 0.109 \text{ L}$

$P_T = 765 \text{ Torr}$

$T = 25°C \rightarrow 25°C + 273.2 = 298.2 \text{ K}$

$R = 62.36 \dfrac{\text{L} \cdot \text{Torr}}{\text{mol} \cdot \text{K}}$

At these conditions, the vapor pressure of water is 23.78 Torr.

$P_T = P_{H_2} + P_{H_2O}$

$P_{H_2} = P_T - P_{H_2O} = 765 \text{ Torr} - 23.78 \text{ Torr} = 741.2 \text{ Torr}$

$PV = nRT \rightarrow n_{H_2} = \dfrac{P_{H_2}V}{RT} = \dfrac{741.2 \text{ Torr} \cdot 0.109 \text{ L}}{62.36 \dfrac{\text{L} \cdot \text{Torr}}{\text{mol} \cdot \text{K}} \cdot 298.2 \text{ K}} = 0.004345 \text{ mol H}_2$

$0.004345 \text{ mol H}_2 \cdot \dfrac{1 \text{ mol Cu}}{1 \text{ mol H}_2} = 0.004345 \text{ mol Cu}$

$0.004345 \text{ mol Cu} \cdot \dfrac{63.55 \text{ g}}{\text{mol}} = 0.276 \text{ g}$

---

48. a.

$V = 2.500 \text{ L}$

$P = 13.11 \text{ atm}$

$T = 22°C \rightarrow 22°C + 273.2 = 295.2 \text{ K}$

$R = 0.08206 \dfrac{\text{L} \cdot \text{Torr}}{\text{mol} \cdot \text{K}}$

$PV = nRT \rightarrow n = \dfrac{PV}{RT} = \dfrac{13.11 \text{ atm} \cdot 2.500 \text{ L}}{0.08206 \dfrac{\text{L} \cdot \text{Torr}}{\text{mol} \cdot \text{K}} \cdot 295.2 \text{ K}} = 1.3530 \text{ mol}$

Rounding to 4 sig digs we have 1.353 mol.

---

48. b.

$1.3530 \text{ mol O}_2 \cdot \dfrac{31.9988 \text{ g}}{\text{mol}} = 43.29 \text{ g}$

---

49. a.

$Fe(s) + CuSO_4(aq) \rightarrow FeSO_4(aq) + Cu(s)$

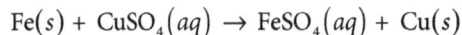

$25.00 \text{ g Fe} \cdot \dfrac{\text{mol}}{55.847 \text{ g}} = 0.44765 \text{ mol Fe}$

$35.8 \text{ g CuSO}_4 \cdot \dfrac{\text{mol}}{159.610 \text{ g}} = 0.2243 \text{ mol CuSO}_4$

Since the mole ratio is 1:1, $CuSO_4$ is the limiting reactant.

---

49. b.

$$0.2243 \text{ mol CuSO}_4 \cdot \frac{1 \text{ mol FeSO}_4}{1 \text{ mol CuSO}_4} = 0.2243 \text{ mol FeSO}_4$$

$$0.2243 \text{ mol FeSO}_4 \cdot \frac{151.911 \text{ g}}{\text{mol}} = 34.1 \text{ g}$$

49. c.

$$0.4477 \text{ mol Fe} - 0.2243 \text{ mol Fe} = 0.2234 \text{ mol Fe}$$

$$0.2234 \text{ mol Fe} \cdot \frac{55.847 \text{ g}}{\text{mol}} = 12.5 \text{ g}$$

50.

$$m = 8.755 \text{ g}$$

$$V = 2.00 \text{ L}$$

$$P = 1.05 \text{ bar} \cdot \frac{100 \text{ kPa}}{1 \text{ bar}} = 105 \text{ kPa}$$

$$T = 30.0°C \rightarrow 30.0°C + 273.2 = 303.2 \text{ K}$$

$$R = 8.314 \frac{\text{L} \cdot \text{kPa}}{\text{mol} \cdot \text{K}}$$

$$PV = nRT \rightarrow n = \frac{PV}{RT} = \frac{105 \text{ kPa} \cdot 2.00 \text{ L}}{8.314 \dfrac{\text{L} \cdot \text{kPa}}{\text{mol} \cdot \text{K}} \cdot 303.2} = 0.08331 \text{ mol}$$

$$M = \frac{8.755 \text{ g}}{0.08331 \text{ mol}} = 105 \frac{\text{g}}{\text{mol}}$$

51.

$$\lambda = 0.22 \text{ nm} \cdot \frac{1 \text{ m}}{10^9 \text{ nm}} = 2.2 \times 10^{-10} \text{ m}$$

$$E = \frac{hv}{\lambda} = \frac{6.626 \times 10^{-34} \text{ J} \cdot \text{s} \cdot 2.9979 \times 10^8 \dfrac{\text{m}}{\text{s}}}{2.2 \times 10^{-10} \text{ m}} = 9.03 \times 10^{-16} \text{ J}$$

$$9.03 \times 10^{-16} \text{ J} \cdot \frac{1 \text{ eV}}{1.602 \times 10^{-19} \text{ J}} = 5600 \text{ eV}$$

52.

$$\text{effusion rate } x = \frac{\text{effusion rate CH}_4}{1.91}$$

$$\frac{\text{effusion rate CH}_4}{\text{effusion rate } x} = 1.91 = \frac{\sqrt{M_x}}{\sqrt{M_{CH_4}}} = \frac{\sqrt{M_x}}{\sqrt{16.04 \ \frac{g}{mol}}}$$

$$\sqrt{M_x} = 1.91 \cdot \sqrt{16.04 \ \frac{g}{mol}}$$

$$M_x = 3.648 \cdot 16.04 \ \frac{g}{mol} = 58.5 \ \frac{g}{mol}$$

Chapter 10

10. a.

$$\frac{1}{2}N_2 + \frac{1}{2}O_2 \rightarrow NO \qquad\qquad \Delta H_f^\circ = 91.3 \text{ kJ/mol}$$

$$\frac{1}{2}N_2 + O_2 \rightarrow NO_2 \qquad\qquad \Delta H_f^\circ = 33.2 \text{ kJ/mol}$$

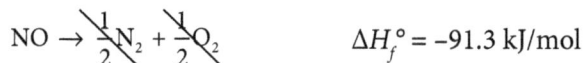 $\qquad\qquad \Delta H_f^\circ = -91.3 \text{ kJ/mol}$

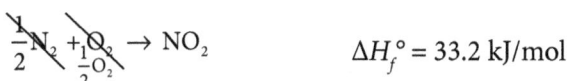 $\qquad\qquad \Delta H_f^\circ = 33.2 \text{ kJ/mol}$

$$NO + \frac{1}{2}O_2 \rightarrow NO_2 \qquad\qquad \Delta H = -58.1 \text{ kJ/mol}$$

10. b.

$$2Fe + \frac{3}{2}O_2 \rightarrow Fe_2O_3 \qquad\qquad \Delta H_f^\circ = -824.2 \text{ kJ/mol}$$

$$C + \frac{1}{2}O_2 \rightarrow CO \qquad\qquad \Delta H_f^\circ = -110.5 \text{ kJ/mol}$$

$$C + O_2 \rightarrow CO_2 \qquad\qquad \Delta H_f^\circ = -393.5 \text{ kJ/mol}$$

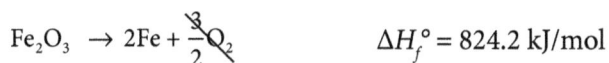 $\qquad\qquad \Delta H_f^\circ = 824.2 \text{ kJ/mol}$

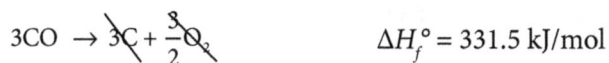 $\qquad\qquad \Delta H_f^\circ = 331.5 \text{ kJ/mol}$

 $\qquad\qquad \Delta H_f^\circ = -1180.5 \text{ kJ/mol}$

$$Fe_2O_3 + 3CO \rightarrow 2Fe + 3CO_2 \qquad\qquad \Delta H = -24.8 \text{ kJ/mol}$$

10. c.

$$Ca + O_2 + H_2 \rightarrow Ca(OH)_2 \qquad \Delta H_f^\circ = -985.2 \text{ kJ/mol}$$

$$Ca + \frac{1}{2}O_2 \rightarrow CaO \qquad \Delta H_f^\circ = -634.9 \text{ kJ/mol}$$

$$H_2 + \frac{1}{2}O_2 \rightarrow H_2O \qquad \Delta H_f^\circ = -241.8 \text{ kJ/mol}$$

$$Ca(OH)_2 \rightarrow \cancel{Ca} + \cancel{O_2} + \cancel{H_2} \qquad \Delta H_f^\circ = 985.2 \text{ kJ/mol}$$

$$\cancel{Ca} + \cancel{\tfrac{1}{2}O_2} \rightarrow CaO \qquad \Delta H_f^\circ = -634.9 \text{ kJ/mol}$$

$$\cancel{H_2} + \cancel{\tfrac{1}{2}O_2} \rightarrow H_2O \qquad \Delta H_f^\circ = -241.8 \text{ kJ/mol}$$

$$Ca(OH)_2 \rightarrow CaO + H_2O \qquad \Delta H = 108.5 \text{ kJ/mol}$$

10. d.

$$Ca + C + \frac{3}{2}O_2 \rightarrow CaCO_3 \qquad \Delta H_f^\circ = -1207.6 \text{ kJ/mol}$$

$$Ca + \frac{1}{2}O_2 \rightarrow CaO \qquad \Delta H_f^\circ = -634.9 \text{ kJ/mol}$$

$$C + O_2 \rightarrow CO_2 \qquad \Delta H_f^\circ = -393.5 \text{ kJ/mol}$$

$$CaCO_3 \rightarrow \cancel{Ca} + \cancel{C} + \frac{3}{2}\cancel{O_2} \qquad \Delta H_f^\circ = 1207.6 \text{ kJ/mol}$$

$$\cancel{Ca} + \cancel{\tfrac{1}{2}O_2} \rightarrow CaO \qquad \Delta H_f^\circ = -634.9 \text{ kJ/mol}$$

$$\cancel{C} + \cancel{O_2} \rightarrow CO_2 \qquad \Delta H_f^\circ = -393.5 \text{ kJ/mol}$$

$$CaCO_3 \rightarrow CaO + CO_2 \qquad \Delta H = 179.2 \text{ kJ/mol}$$

10. e.

$$Mg + O_2 + H_2 \rightarrow Mg(OH)_2 \qquad \Delta H_f^\circ = -924.5 \text{ kJ/mol}$$

$$Mg + \frac{1}{2}O_2 \rightarrow MgO \qquad \Delta H_f^\circ = -601.6 \text{ kJ/mol}$$

$$H_2 + \frac{1}{2}O_2 \rightarrow H_2O \qquad \Delta H_f^\circ = -285.8 \text{ kJ/mol}$$

$$Mg(OH)_2 \rightarrow Mg + O_2 + H_2 \qquad \Delta H_f^\circ = 924.5 \text{ kJ/mol}$$

$$\cancel{Mg} + \cancel{\tfrac{1}{2}O_2} \rightarrow MgO \qquad \Delta H_f^\circ = -601.6 \text{ kJ/mol}$$

$$\cancel{H_2} + \cancel{\tfrac{1}{2}O_2} \rightarrow H_2O \qquad \Delta H_f^\circ = -285.8 \text{ kJ/mol}$$

$$Mg(OH)_2 \rightarrow MgO + H_2O \qquad \Delta H = 37.1 \text{ kJ/mol}$$

11. a.

Reactants		Products	
	ΔH_f° (kJ/mol)		ΔH_f° (kJ/mol)
NO(g)	91.3	NO$_2$(g)	33.2
½O$_2$(g)	0		
Sum (Σ)	91.3		33.2

$\Delta H^\circ = \Sigma \Delta H^\circ_{f,\,products} - \Sigma \Delta H^\circ_{f,\,reactants} = 33.2 \text{ kJ/mol} - 91.3 \text{ kJ/mol} = -58.1 \text{ kJ/mol}$

11. b.

Reactants		Products	
	ΔH_f° (kJ/mol)		ΔH_f° (kJ/mol)
Fe$_2$O$_3$(s)	−824.2	2Fe(s)	0
3CO(g)	3(−110.5) = −331.5	3CO$_2$(g)	3(−393.5) = −1180.5
Sum (Σ)	−1155.7		−1180.5

$\Delta H^\circ = \Sigma \Delta H^\circ_{f,\,products} - \Sigma \Delta H^\circ_{f,\,reactants} = -1180.5 \text{ kJ/mol} - (-1155.7 \text{ kJ/mol}) = -24.8 \text{ kJ/mol}$

11. c.

	Reactants		Products	
	ΔH_f° (kJ/mol)			ΔH_f° (kJ/mol)
$Ca(OH)_2(s)$	-985.2	$CaO(s)$		-634.9
		$H_2O(g)$		-241.8
Sum (Σ)	-985.2			-876.7

$\Delta H° = \Sigma\Delta H°_{f,\,products} - \Sigma\Delta H°_{f,\,reactants} = -876.7 \text{ kJ/mol} - (-985.2 \text{ kJ/mol}) = 108.5 \text{ kJ/mol}$

11. d.

	Reactants		Products	
	ΔH_f° (kJ/mol)			ΔH_f° (kJ/mol)
$CaCO_3(s)$	-1207.6	$CaO(s)$		-634.9
		$CO_2(g)$		-393.5
Sum (Σ)	-1207.6			-1028.4

$\Delta H° = \Sigma\Delta H°_{f,\,products} - \Sigma\Delta H°_{f,\,reactants} = -1028.4 \text{ kJ/mol} - (-1207.6 \text{ kJ/mol}) = 179.2 \text{ kJ/mol}$

11. e.

	Reactants		Products	
	ΔH_f° (kJ/mol)			ΔH_f° (kJ/mol)
$Mg(OH)_2(s)$	-924.5	$MgO(s)$		-601.6
		$H_2O(l)$		-285.8
Sum (Σ)	-924.5			-887.4

$\Delta H° = \Sigma\Delta H°_{f,\,products} - \Sigma\Delta H°_{f,\,reactants} = -887.4 \text{ kJ/mol} - (-924.5 \text{ kJ/mol}) = 37.1 \text{ kJ/mol}$

12. a.

Formation equation: $2C + 2H_2 \rightarrow C_2H_4$

$C + O_2 \rightarrow CO_2$	$\Delta H_c^\circ = -393.5$ kJ/mol
$H_2 + \dfrac{1}{2}O_2 \rightarrow H_2O$	$\Delta H_c^\circ = -285.8$ kJ/mol
$C_2H_4 + 3O_2 \rightarrow 2CO_2 + 2H_2O$	$\Delta H_c^\circ = -1411.2$ kJ/mol

$2C + 2O_2 \rightarrow 2CO_2$	$\Delta H = 2(-393.5)$ kJ/mol
$2H_2 + O_2 \rightarrow 2H_2O$	$\Delta H = 2(-285.8)$ kJ/mol
$2CO_2 + 2H_2O \rightarrow C_2H_4 + 3O_2$	$\Delta H = 1411.2$ kJ/mol
$2C + 2H_2 \rightarrow C_2H_4$	$\Delta H_f^\circ = 52.6$ kJ/mol

12. b.

Formation equation: $C + H_2 + O_2 \rightarrow CH_2O_2$

$C + O_2 \rightarrow CO_2$	$\Delta H_c^\circ = -393.5$ kJ/mol
$H_2 + \dfrac{1}{2}O_2 \rightarrow H_2O$	$\Delta H_c^\circ = -285.8$ kJ/mol
$CH_2O_2 + \dfrac{1}{2}O_2 \rightarrow CO_2 + H_2O$	$\Delta H_c^\circ = -254.6$ kJ/mol

$C + O_2 \rightarrow CO_2$	$\Delta H = -393.5$ kJ/mol
$H_2 + \dfrac{1}{2}O_2 \rightarrow H_2O$	$\Delta H = -285.8$ kJ/mol
$CO_2 + H_2O \rightarrow CH_2O_2 + \dfrac{1}{2}O_2$	$\Delta H = 254.6$ kJ/mol
$C + H_2 + O_2 \rightarrow CH_2O_2$	$\Delta H_f^\circ = -424.7$ kJ/mol

12. c.

Formation equation: $6C + 3H_2 \rightarrow C_6H_6$

$$C + O_2 \rightarrow CO_2 \qquad\qquad \Delta H_c^\circ = -393.5 \text{ kJ/mol}$$

$$H_2 + \frac{1}{2}O_2 \rightarrow H_2O \qquad\qquad \Delta H_c^\circ = -285.8 \text{ kJ/mol}$$

$$C_6H_6 + \frac{15}{2}O_2 \rightarrow 6CO_2 + 3H_2O \qquad\qquad \Delta H_c^\circ = -3267.6 \text{ kJ/mol}$$

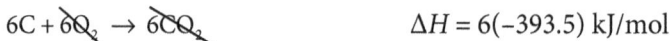 $\qquad\qquad \Delta H = 6(-393.5) \text{ kJ/mol}$

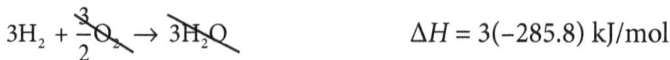 $\qquad\qquad \Delta H = 3(-285.8) \text{ kJ/mol}$

 $\qquad\qquad \Delta H = 3267.6 \text{ kJ/mol}$

$$6C + 3H_2 \rightarrow C_6H_6 \qquad\qquad \Delta H_f^\circ = 49.2 \text{ kJ/mol}$$

14. a.

Combustion equation: $CH_3OH + \frac{3}{2}O_2 \rightarrow CO_2 + 2H_2O$

$$C + 2H_2 + \frac{1}{2}O_2 \rightarrow CH_3OH \qquad\qquad \Delta H_f^\circ = -239.2 \text{ kJ/mol}$$

$$C + O_2 \rightarrow CO_2 \qquad\qquad \Delta H_f^\circ = -393.5 \text{ kJ/mol}$$

$$H_2 + \frac{1}{2}O_2 \rightarrow H_2O \qquad\qquad \Delta H_f^\circ = -285.8 \text{ kJ/mol}$$

 $\qquad\qquad \Delta H = 239.2 \text{ kJ/mol}$

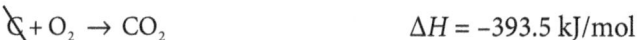 $\qquad\qquad \Delta H = -393.5 \text{ kJ/mol}$

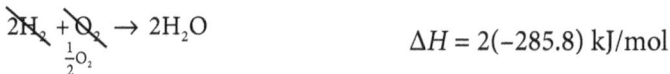 $\qquad\qquad \Delta H = 2(-285.8) \text{ kJ/mol}$

$$CH_3OH + \frac{3}{2}O_2 \rightarrow CO_2 + 2H_2O \qquad\qquad \Delta H_c^\circ = -725.9 \text{ kJ/mol}$$

13. b.

Combustion equation: $C_2H_5OH + 3O_2 \rightarrow 2CO_2 + 3H_2O$

$2C + 3H_2 + \frac{1}{2}O_2 \rightarrow C_2H_5OH$	$\Delta H_f^\circ = -277.6$ kJ/mol
$C + O_2 \rightarrow CO_2$	$\Delta H_f^\circ = -393.5$ kJ/mol
$H_2 + \frac{1}{2}O_2 \rightarrow H_2O$	$\Delta H_f^\circ = -285.8$ kJ/mol

$C_2H_5OH \rightarrow \cancel{2C} + \cancel{3H_2} + \cancel{\frac{1}{2}O_2}$	$\Delta H = 277.6$ kJ/mol
$\cancel{2C} + \cancel{\frac{3}{2}O_2} \rightarrow 2CO_2$	$\Delta H = 2(-393.5)$ kJ/mol
$\cancel{3H_2} + \frac{3}{2}O_2 \rightarrow 3H_2O$	$\Delta H = 3(-285.8)$ kJ/mol
$C_2H_5OH + 3O_2 \rightarrow 2CO_2 + 3H_2O$	$\Delta H_c^\circ = -1366.8$ kJ/mol

---

19. a.

$\Delta H^\circ = 182.6$ kJ

$\Delta S^\circ = 24.8 \dfrac{J}{K} \cdot \dfrac{kJ}{1000\ J} = 0.0248 \dfrac{kJ}{K}$

$T = 25°C \rightarrow 25°C + 273.2 = 298.2$ K

$\Delta G^\circ = \Delta H^\circ - T\Delta S^\circ = 182.6$ kJ $- 298.2$ K $\cdot 0.0248 \dfrac{kJ}{K} = 182.6$ kJ $- 7.40$ kJ $= 175.2$ kJ

Since ΔG° is positive, this reaction does not occur spontaneously.

19. b.

$\Delta H^\circ = -393.5$ kJ

$\Delta S^\circ = 2.70 \dfrac{J}{K} \cdot \dfrac{kJ}{1000\ J} = 0.00270 \dfrac{kJ}{K}$

$T = 25°C \rightarrow 25°C + 273.2 = 298.2$ K

$\Delta G^\circ = \Delta H^\circ - T\Delta S^\circ = -393.5$ kJ $- 298.2$ K $\cdot 0.00270 \dfrac{kJ}{K} = -393.5$ kJ $- 0.805$ kJ $= -394.3$ kJ

Since ΔG° is negative, this reaction does occur spontaneously.

20. a.

$$\Delta H = 108 \ \frac{kJ}{mol}$$

$$\Delta S = 0.0375 \ \frac{kJ}{mol \cdot K}$$

$$T = 292 \ K$$

$$\Delta G = \Delta H - T\Delta S = 108 \ \frac{kJ}{mol} - 292 \ K \cdot 0.0375 \ \frac{kJ}{mol \cdot K} = 108 \ \frac{kJ}{mol} - 11.0 \ \frac{kJ}{mol} = 97 \ \frac{kJ}{mol}$$

(not spontaneous)

---

20. b.

$$\Delta H = -93.1 \ \frac{kJ}{mol}$$

$$\Delta S = 0.105 \ \frac{kJ}{mol \cdot K}$$

$$T = 135°C \rightarrow 135°C + 273.2 = 408 \ K$$

$$\Delta G = \Delta H - T\Delta S = -93.1 \ \frac{kJ}{mol} - 408 \ K \cdot 0.105 \ \frac{kJ}{mol \cdot K} = -93.1 \ \frac{kJ}{mol} - 42.8 \ \frac{kJ}{mol} = -135.9 \ \frac{kJ}{mol}$$

(spontaneous)

---

20. c.

$$\Delta H = -278 \ \frac{kJ}{mol}$$

$$\Delta S = 0.375 \ \frac{kJ}{mol \cdot K}$$

$$T = 775 \ K$$

$$\Delta G = \Delta H - T\Delta S = -278 \ \frac{kJ}{mol} - 775 \ K \cdot 0.375 \ \frac{kJ}{mol \cdot K} = -278 \ \frac{kJ}{mol} - 291 \ \frac{kJ}{mol} = -569 \ \frac{kJ}{mol}$$

(spontaneous)

21.

	Reactants			Products		Σproducts − Σreactants
	N_2	$3H_2$	Sum (Σ)	$2NH_3$	Sum (Σ)	
ΔH_f° kJ/mol	0	0	0	2(−45.9)	−91.8	−91.8
ΔS° kJ/(mol·K)	0.1916	3(0.1307)	0.5837	2(0.1928)	0.3856	−0.1981

$T = 25°C \rightarrow 25°C + 273 = 298$ K

$$\Delta G^\circ = \Delta H^\circ - T\Delta S^\circ = -91.8 \ \frac{kJ}{mol} - 298 \ K \cdot \left(-0.1981 \ \frac{kJ}{mol \cdot K} \right) = -32.7 \ \frac{kJ}{mol}$$

$T = 500°C \rightarrow 500°C + 273 = 773$ K

$$\Delta G^\circ = \Delta H^\circ - T\Delta S^\circ = -91.8 \ \frac{kJ}{mol} - 773 \ K \cdot \left(-0.1981 \ \frac{kJ}{mol \cdot K} \right) = 61.3 \ \frac{kJ}{mol}$$

22. a.

	Reactants			Products		
	$6Cl_2$	$2Fe_2O_3$	Sum (Σ)	$4FeCl_3$	$3O_2$	Sum (Σ)
ΔG_f° kJ/mol		2(−742.2)	−1484.4	4(−334.0)		−1336.0

$$\Delta G^\circ = \Sigma\Delta G^\circ_{f,\text{products}} - \Sigma\Delta G^\circ_{f,\text{reactants}} = -1336.0 \ \frac{kJ}{mol} - \left(-1484.4 \ \frac{kJ}{mol} \right) = 148.4 \ \frac{kJ}{mol}$$

	Reactants			Products			Σprods − Σreacts
	$6Cl_2$	$2Fe_2O_3$	Sum (Σ)	$4FeCl_3$	$3O_2$	Sum (Σ)	
ΔH_f° kJ/mol	0	2(−824.2)	−1648.4	4(−399.5)	0	−1598.0	$\Delta H^\circ =$ 50.4 kJ/mol
ΔS° kJ/(mol·K)	6(0.2231)	2(0.0874)	1.5134	4(0.1423)	3(0.2052)	1.1848	$\Delta S^\circ =$ −0.3286 kJ/ (mol·K)

$T = 25°C \rightarrow 25°C + 273 = 298$ K

$$\Delta G^\circ = \Delta H^\circ - T\Delta S^\circ = 50.4 \ \frac{kJ}{mol} - 298 \ K \cdot \left(-0.3286 \ \frac{kJ}{mol \cdot K} \right) = 148.3 \ \frac{kJ}{mol}$$

22. b.

	Reactants			Products	
	H_2	F_2	Sum (Σ)	2HF	Sum (Σ)
ΔG_f° kJ/mol				2(–275.4)	–550.8

$$\Delta G^\circ = \Sigma \Delta G^\circ_{f,\,products} - \Sigma \Delta G^\circ_{f,\,reactants} = -550.8 \ \frac{kJ}{mol}$$

	Reactants			Products		Σprods – Σreacts
	H_2	F_2	Sum (Σ)	2HF	Sum (Σ)	
ΔH_f° kJ/mol	0	0	0	2(–273.3)	–546.6	$\Delta H^\circ = -546.6$ kJ/mol
ΔS° kJ/(mol·K)	0.1307	0.2028	0.3335	2(0.1738)	0.3476	$\Delta S^\circ = 0.0141$ kJ/(mol·K)

$T = 25°C \rightarrow 25°C + 273 = 298$ K

$$\Delta G^\circ = \Delta H^\circ - T\Delta S^\circ = -546.6 \ \frac{kJ}{mol} - 298 \ K \cdot \left(0.0141 \ \frac{kJ}{mol \cdot K} \right) = -550.8 \ \frac{kJ}{mol}$$

22. c.

	Reactants			Products	
	NO_2	N_2O	Sum (Σ)	3NO	Sum (Σ)
ΔG_f° kJ/mol	51.3	103.7	155.0	3(87.6)	262.8

$$\Delta G^\circ = \Sigma \Delta G^\circ_{f,\,products} - \Sigma \Delta G^\circ_{f,\,reactants} = 262.8 \ \frac{kJ}{mol} - 155.0 \ \frac{kJ}{mol} = 107.8 \ \frac{kJ}{mol}$$

	Reactants			Products		Σprods – Σreacts
	NO_2	N_2O	Sum (Σ)	3NO	Sum (Σ)	
ΔH_f° kJ/mol	33.2	81.6	114.8	3(91.3)	273.9	$\Delta H^\circ = 159.1$ kJ/mol
ΔS° kJ/(mol·K)	0.2401	0.2200	0.4601	3(0.2108)	0.6324	$\Delta S^\circ = 0.1723$ kJ/(mol·K)

$T = 25°C \rightarrow 25°C + 273 = 298$ K

$$\Delta G^\circ = \Delta H^\circ - T\Delta S^\circ = 159.1 \ \frac{kJ}{mol} - 298 \ K \cdot \left(0.1723 \ \frac{kJ}{mol \cdot K} \right) = 107.8 \ \frac{kJ}{mol}$$

22. d.

	Reactants			Products		
	SO_2	$2H_2$	Sum (Σ)	S	$2H_2O$	Sum (Σ)
ΔG_f° kJ/mol	−300.1		−300.1		2(−228.6)	−457.2

$$\Delta G^\circ = \Sigma\Delta G^\circ_{f,\text{products}} - \Sigma\Delta G^\circ_{f,\text{reactants}} = -457.2\ \frac{kJ}{mol} - \left(-300.1\ \frac{kJ}{mol}\right) = -157.1\ \frac{kJ}{mol}$$

	Reactants			Products			Σprods − Σreacts
	SO_2	$2H_2$	Sum (Σ)	S	$2H_2O$	Sum (Σ)	
ΔH_f° kJ/mol	−296.8	0	−296.8	0	2(−241.8)	−483.6	$\Delta H^\circ =$ −186.8 kJ/mol
ΔS° kJ/(mol·K)	0.2482	2(0.1307)	0.5096	0.0321	2(0.1888)	0.4097	$\Delta S^\circ =$ −0.0999 kJ/ (mol·K)

$$T = 25°C \rightarrow 25°C + 273 = 298\ K$$

$$\Delta G^\circ = \Delta H^\circ - T\Delta S^\circ = -186.8\ \frac{kJ}{mol} - 298\ K\cdot\left(-0.0999\ \frac{kJ}{mol\cdot K}\right) = -157.0\ \frac{kJ}{mol}$$

22. e.

	Reactants			Products		
	$MgCl_2$	H_2O	Sum (Σ)	MgO	2HCl	Sum (Σ)
ΔG_f° kJ/mol	−591.8	−237.1	−828.9	−569.3	2(−95.3)	−759.9

$$\Delta G^\circ = \Sigma\Delta G^\circ_{f,\text{products}} - \Sigma\Delta G^\circ_{f,\text{reactants}} = -759.9\ \frac{kJ}{mol} - \left(-828.9\ \frac{kJ}{mol}\right) = 69.0\ \frac{kJ}{mol}$$

	Reactants			Products			Σprods − Σreacts
	$MgCl_2$	H_2O	Sum (Σ)	MgO	2HCl	Sum (Σ)	
ΔH_f° kJ/mol	−641.3	−285.8	−927.1	−601.6	2(−92.3)	−786.2	$\Delta H^\circ =$ 140.9 kJ/mol
ΔS° kJ/(mol·K)	0.0896	0.0700	0.1596	0.0270	2(0.1869)	0.4008	$\Delta S^\circ =$ 0.2412 kJ/ (mol·K)

$$T = 25°C \rightarrow 25°C + 273 = 298\ K$$

$$\Delta G^\circ = \Delta H^\circ - T\Delta S^\circ = 140.9\ \frac{kJ}{mol} - 298\ K\cdot 0.2412\ \frac{kJ}{mol\cdot K} = 69.0\ \frac{kJ}{mol}$$

31. b.

Using data from Trial 1:

$$k = \frac{R}{A^2 \cdot B} = \frac{1.15 \times 10^{-4} \dfrac{M}{s}}{(0.10 \ M)^2 \cdot 0.20 \ M} = 0.058 \ s^{-1} M^{-2}$$

---

31. c.

$$R = kA^2 B = \frac{0.058}{s \cdot M^2} \cdot (0.25 \ M)^2 \cdot 0.25 \ M = 0.00091 \ \frac{M}{s}$$

32.

$m = 3.55 \ g$

$P = 100 \ kPa$

$T = 273.15 \ K$

$R = 8.314 \ \dfrac{L \cdot kPa}{mol \cdot K}$

$n = 3.55 \ g \cdot \dfrac{mol}{80.064 \ g} = 0.04434 \ mol$

$$PV = nRT \rightarrow V = \frac{nRT}{P} = \frac{0.04434 \ mol \cdot 8.314 \ \dfrac{L \cdot kPa}{mol \cdot K} \cdot 273.15 \ K}{100 \ kPa} = 1.01 \ L$$

33.

$4Al + 3O_2 \rightarrow 2Al_2O_3$

$0.046 \ mol \ Al \cdot \dfrac{3 \ mol \ O_2}{4 \ mol \ Al} = 0.035 \ mol \ O_2$

This much O_2 is not available, so O_2 is the limiting reactant.

$0.028 \ mol \ O_2 \cdot \dfrac{2 \ mol \ Al_2O_3}{3 \ mol \ O_2} = 0.0187 \ mol \ Al_2O_3$

$0.0187 \ mol \ Al_2O_3 \cdot \dfrac{101.96 \ g}{mol} = 1.9 \ g$

34.

$3F_2 + 3H_2O \rightarrow 6HF + O_3$

$5.500 \ L \ F_2 \cdot \dfrac{6 \ L \ HF}{3 \ L \ F_2} = 11.00 \ L \ HF$

$5.500 \ L \ F_2 \cdot \dfrac{1 \ L \ O_2}{3 \ L \ F_2} = 1.833 \ L \ O_2$

35.

$$132.7 \text{ g } C_6H_8O_7 \cdot \frac{\text{mol}}{192.125 \text{ g}} = 0.69070 \text{ mol}$$

$$\frac{0.69070 \text{ mol}}{2500.0 \text{ mL}} \cdot \frac{1000 \text{ mL}}{1 \text{ L}} = 0.2763 \text{ } M$$

36.

$$pH = 9.81$$

$$pH + pOH = 14.0$$

$$pOH = 14.0 - pH = 14.0 - 9.81 = 4.2$$

$$[H_3O^+] = 10^{-pH} = 10^{-9.81} = 1.5 \times 10^{-10} \text{ } M$$

$$[H_3O^+] \cdot [OH^-] = 1.0 \times 10^{-14} \text{ } M^2$$

$$[OH^-] = \frac{1.0 \times 10^{-14} \text{ } M^2}{[H_3O^+]} = \frac{1.0 \times 10^{-14} \text{ } M^2}{1.5 \times 10^{-10} \text{ } M} = 6.7 \times 10^{-5} \text{ } M$$

37.

$$K_f = 1.86 \frac{°C}{m}$$

$$K_b = 0.513 \frac{°C}{m}$$

$$175.00 \text{ g KBr} \cdot \frac{\text{mol}}{119.002 \text{ g}} = 1.4706 \text{ mol KBr}$$

$$1.4706 \text{ mol KBr} \cdot \frac{2 \text{ mol ions}}{1 \text{ mol KBr}} = 2.9411 \text{ mol ions}$$

$$450.0 \text{ g } H_2O \cdot \frac{1 \text{ kg}}{1000 \text{ g}} = 0.4500 \text{ kg } H_2O$$

$$m = \frac{2.9411 \text{ mol ions}}{0.4500 \text{ kg } H_2O} = 6.536 \text{ } m$$

$$\Delta T_f = K_f m = 1.86 \frac{°C}{m} \cdot 6.536 \text{ } m = 12.2°C$$

$$T_f = 0.00°C - 12.2°C = -12.2°C$$

$$\Delta T_b = K_b m = 0.513 \frac{°C}{m} \cdot 6.536 \text{ } m = 3.35°C$$

$$T_b = 99.974°C + 3.35°C = 103.32°C$$

38.

$$17.00 \text{ mL} \cdot \frac{1 \text{ L}}{1000 \text{ mL}} \cdot \frac{0.650 \text{ mol}}{L} = 0.01105 \text{ mol NaOH}$$

$$0.01105 \text{ mol NaOH} \rightarrow 0.01105 \text{ mol } CH_3COOH$$

$$\frac{0.01105 \text{ mol } CH_3COOH}{18.50 \text{ mL}} \cdot \frac{1000 \text{ mL}}{L} = 0.597 \text{ } M$$

Chapter 11

7.

Note that in the first edition text the equation is 6(a) is not correctly balanced. The correct equation is:

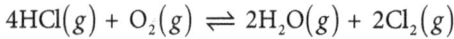

$$4HCl(g) + O_2(g) \rightleftharpoons 2H_2O(g) + 2Cl_2(g)$$

$$[HCl] = 2.4 \times 10^{-3} \; M$$

$$[O_2] = 7.6 \times 10^{-8} \; M$$

$$[H_2O] = 1.2 \times 10^{-1} \; M$$

$$[Cl_2] = 1.2 \times 10^{-1} \; M$$

$$K = \frac{[H_2O]^2 [Cl_2]^2}{[HCl]^4 [O_2]} = \frac{\left(1.2 \times 10^{-1} \; M\right)^2 \left(1.2 \times 10^{-1} \; M\right)^2}{\left(2.4 \times 10^{-3} \; M\right)^4 \left(7.6 \times 10^{-8} \; M\right)} = 8.2 \times 10^{-13}$$

8.

$$N_2 + 3H_2 \rightarrow 2NH_3$$

$$K = 6.59 \times 10^{-3}$$

$$[H_2] = 2.61 \times 10^{-3} \; M$$

$$[NH_3] = 1.17 \times 10^{-4} \; M$$

$$K = \frac{[NH_3]^2}{[N_2][H_2]^3} \rightarrow [N_2] = \frac{[NH_3]^2}{K \cdot [H_2]^3} = \frac{\left(1.17 \times 10^{-4} \; M\right)^2}{6.59 \times 10^{-3} \cdot \left(2.61 \times 10^{-3} \; M\right)^3} = 117 \; M$$

---

9. a.

$$K = 2.19 \times 10^{-10}$$

$$[COCl_2] = 2.00 \times 10^{-3} \; M$$

$$[CO] = 3.3 \times 10^{-6} \; M$$

$$[Cl_2] = 6.62 \times 10^{-6} \; M$$

$$\frac{[CO][Cl_2]}{[COCl_2]} = \frac{3.3 \times 10^{-6} \; M \cdot 6.62 \times 10^{-6} \; M}{2.00 \times 10^{-3} \; M} = 1.09 \times 10^{-8}$$

$$1.09 \times 10^{-8} > K$$

Since the product concentrations are higher than the reactant concentrations relative to their equilibrium ratio K, the reaction must shift to the left to achieve equilibrium.

---

9. b.

$K = 2.19 \times 10^{-10}$

$[COCl_2] = 4.50 \times 10^{-2} \ M$

$[CO] = 1.1 \times 10^{-7} \ M$

$[Cl_2] = 2.25 \times 10^{-6} \ M$

$$\frac{[CO][Cl_2]}{[COCl_2]} = \frac{1.1 \times 10^{-7} \ M \cdot 2.25 \times 10^{-6} \ M}{4.50 \times 10^{-2} \ M} = 5.5 \times 10^{-12}$$

$5.5 \times 10^{-12} < K$

Since the reactant concentrations are lower than the product concentrations relative to their equilibrium ratio K, the reaction must shift to the right to achieve equilibrium.

---

9. c.

$K = 2.19 \times 10^{-10}$

$[COCl_2] = 0.0100 \ M$

$[CO] = 1.48 \times 10^{-6} \ M$

$[Cl_2] = 1.48 \times 10^{-6} \ M$

$$\frac{[CO][Cl_2]}{[COCl_2]} = \frac{1.48 \times 10^{-6} \ M \cdot 1.48 \times 10^{-6} \ M}{0.0100 \ M} = 2.19 \times 10^{-10}$$

$5.5 \times 10^{-12} = K$

The reaction is at equilibrium.

10.

$K = 1.7 \times 10^{13}$

$[N_2O] = 0.0029 \ M$

$[O_2] = 0.0023 \ M$

$$K = \frac{[NO]^2}{[N_2O][O_2]^{\frac{1}{2}}}$$

$[NO]^2 = K[N_2O][O_2]^{\frac{1}{2}}$

$$[NO] = \left(K[N_2O][O_2]^{\frac{1}{2}} \right)^{\frac{1}{2}} = \left(1.7 \times 10^{13} \cdot 0.0029 \ M \cdot (0.0023 \ M)^{\frac{1}{2}} \right)^{\frac{1}{2}} = 4.9 \times 10^4 \ M$$

---

12. b.

$H_2 + I_2 \rightarrow 2HI$

$K_f = 40.0$

$$K_f = \frac{[HI]^2}{[H_2][I_2]} = 40.0$$

Reverse reaction:

$$K_r = \frac{[H_2][I_2]}{[HI]^2} = \frac{1}{K_f} = \frac{1}{40.0} = 0.0250$$

17.

$pH = 2.44$

$[HA] = 0.10 \ M$

$[H_3O^+] = 10^{-pH} = 10^{-2.44} = 0.0036 \ M$

$$K_a = \frac{[H_3O^+][A^-]}{[HA]} = \frac{0.0036 \cdot 0.0036}{0.10} = 1.3 \times 10^{-4}$$

18.

$pH = 1.85$

$[HA] = 0.0500 \ M$

$[H_3O^+] = 10^{-pH} = 10^{-1.85} = 0.014 \ M$

$$K_a = \frac{[H_3O^+][A^-]}{[HA]} = \frac{0.014 \cdot 0.014}{0.0500} = 0.0039$$

19.

$[HA] = 0.868 \cdot 0.100 \ M = 0.0868 \ M$

$[H_3O^+] = [A^-] = 0.132 \cdot 0.100 \ M = 0.0132 \ M$

$pH = -\log[H_3O^+] = -\log 0.0132 = 1.879$

$$K_a = \frac{[H_3O^+][A^-]}{[HA]} = \frac{0.0132 \cdot 0.0132}{0.0868} = 0.00201$$

20.

$[ClCH_2COOH] = 0.890 \cdot 0.100 \ M = 0.0890 \ M$

$[H_3O^+] = [ClCH_2COO^-] = 0.110 \cdot 0.100 \ M = 0.0110 \ M$

$$K_a = \frac{[H_3O^+][ClCH_2COO^-]}{[ClCH_2COOH]} = \frac{0.0110 \cdot 0.0110}{0.0890} = 0.00136$$

21.

$K_a = 1.7 \times 10^{-5}$

$\text{pH} = 2.92$

$\left[H_3O^+ \right] = \left[CH_3COO^- \right] = 10^{-\text{pH}} = 10^{-2.92} = 0.0012$

$K_a = \dfrac{\left[H_3O^+ \right]\left[CH_3COO^- \right]}{\left[CH_3COOH \right]} \rightarrow \left[CH_3COOH \right] = \dfrac{\left[H_3O^+ \right]\left[CH_3COO^- \right]}{K_a} = \dfrac{0.0012 \cdot 0.0012}{1.7 \times 10^{-5}} = 0.085 \ M$

22.

Let P = "proportion ionized"

Let %P = P·100% = "percent ionization"

First concentration:

$M = 0.150 \ M$

$\left[H_3O^+ \right] = \left[CN^- \right] = 0.150 \cdot P$

$\left[HCN \right] = 0.150 \cdot (1-P)$

$K_a = 6.2 \times 10^{-10}$

$K_a = 6.2 \times 10^{-10} = \dfrac{\left[H_3O^+ \right]\left[CN^- \right]}{\left[HCN \right]} = \dfrac{(0.150 \cdot P)(0.150 \cdot P)}{0.150 \cdot (1-P)}$

$6.2 \times 10^{-10} \cdot 0.150 \cdot (1-P) = (0.150 \cdot P)(0.150 \cdot P)$

$9.3 \times 10^{-11} - 9.3 \times 10^{-11} P = 0.0225 P^2$

$0.0225 P^2 + 9.3 \times 10^{-11} P - 9.3 \times 10^{-11} = 0$

$P^2 + 4.13 \times 10^{-9} P - 4.13 \times 10^{-9} = 0$

$P = \dfrac{-4.13 \times 10^{-9} + \sqrt{\left(4.13 \times 10^{-9} \right)^2 - 4 \cdot \left(-4.13 \times 10^{-9} \right)}}{2} = 6.4 \times 10^{-5}$

$\%P = 6.4 \times 10^{-5} \cdot 100\% = 0.0064\%$

Second concentration:

$M = 0.0800 \ M$

$\left[H_3O^+ \right] = \left[CN^- \right] = 0.0800 \cdot P$

$[HCN] = 0.0800 \cdot (1-P)$

$K_a = 6.2 \times 10^{-10}$

$K_a = 6.2 \times 10^{-10} = \dfrac{\left[H_3O^+ \right]\left[CN^- \right]}{[HCN]} = \dfrac{(0.0800 \cdot P)(0.0800 \cdot P)}{0.0800 \cdot (1-P)}$

$6.2 \times 10^{-10} \cdot 0.0800 \cdot (1-P) = (0.0800 \cdot P)(0.0800 \cdot P)$

$4.96 \times 10^{-11} - 4.96 \times 10^{-11} P = 0.0064 P^2$

$0.0064 P^2 + 4.96 \times 10^{-11} P - 4.96 \times 10^{-11} = 0$

$P^2 + 7.75 \times 10^{-9} P - 7.75 \times 10^{-9} = 0$

$P = \dfrac{-7.75 \times 10^{-9} + \sqrt{\left(7.75 \times 10^{-9}\right)^2 - 4 \cdot \left(-7.75 \times 10^{-9}\right)}}{2} = 8.8 \times 10^{-5}$

$\%P = 8.8 \times 10^{-5} \cdot 100\% = 0.0088\%$

Third concentration:

$M = 0.0200 \ M$

$\left[H_3O^+ \right] = \left[CN^- \right] = 0.0200 \cdot P$

$[HCN] = 0.0200 \cdot (1-P)$

$K_a = 6.2 \times 10^{-10}$

$K_a = 6.2 \times 10^{-10} = \dfrac{\left[H_3O^+ \right]\left[CN^- \right]}{[HCN]} = \dfrac{(0.0200 \cdot P)(0.0200 \cdot P)}{0.0200 \cdot (1-P)}$

$6.2 \times 10^{-10} \cdot 0.0200 \cdot (1-P) = (0.0200 \cdot P)(0.0200 \cdot P)$

$1.24 \times 10^{-11} - 1.24 \times 10^{-11} P = 0.00040 P^2$

$0.00040 P^2 + 1.24 \times 10^{-11} P - 1.24 \times 10^{-11} = 0$

$P^2 + 3.1 \times 10^{-8} P - 3.1 \times 10^{-8} = 0$

$P = \dfrac{-3.1 \times 10^{-8} + \sqrt{\left(3.1 \times 10^{-8}\right)^2 - 4 \cdot \left(-3.1 \times 10^{-8}\right)}}{2} = 1.8 \times 10^{-4}$

$\%P = 1.8 \times 10^{-4} \cdot 100\% = 0.00018\%$

23.

In water at 25°C, $[H_3O^+] = [OH^-] = 1.0 \times 10^{-7}$, and $[H_3O^+] \cdot [OH^-] = K_w$. According to equation (11.8), $K_a K_b = K_w$. Since the concentrations of $[H_3O^+]$ and $[OH^-]$ are equal, their dissociation constants are equal. And since their product is K_w, K_b for $[OH^-]$ is 1.0×10^{-7}.

---

25. a.

pH = 11.33

$\left[H_3O^+\right] = 10^{-pH} = 10^{-11.33} = 4.7\times10^{-12}\ M$

$\left[H_3O^+\right]\cdot\left[OH^-\right] = 1.0\times10^{-14}\ M^2$

$\left[OH^-\right] = \dfrac{1.0\times10^{-14}\ M^2}{\left[H_3O^+\right]} = \dfrac{1.0\times10^{-14}\ M^2}{4.7\times10^{-12}\ M} = 2.1\times10^{-3}\ M$

$\left[OH^-\right] = \left[C_{10}H_{15}ONH^+\right]$, so $\left[C_{10}H_{15}ONH^+\right] = 2.1\times10^{-3}\ M$

$\left[C_{10}H_{15}ON\right] \approx 0.035\ M$

---

25. b.

$K_b = \dfrac{\left[H_3O^+\right]\left[C_{10}H_{15}ONH^+\right]}{\left[C_{10}H_{15}ON\right]} = \dfrac{\left(2.1\times10^{-3}\right)\left(2.1\times10^{-3}\right)}{0.035} = 1.3\times10^{-4}$

---

26. c.

$NH_3 : K_b = 1.8\times10^{-5}$

$NH_2OH : K_b = 8.7\times10^{-9}$

$K_a \cdot K_b = K_w$

$K_a = \dfrac{K_w}{K_b}$

For NH_4^+:

$K_a = \dfrac{K_w}{K_b} = \dfrac{1.0\times10^{-14}}{1.8\times10^{-5}} = 5.6\times10^{-10}$

For NH_3OH^+:

$K_a = \dfrac{K_w}{K_b} = \dfrac{1.0\times10^{-14}}{8.7\times10^{-9}} = 1.1\times10^{-6}$

30.

$CaF(s) \rightarrow Ca^{2+}(aq) + 2F^-(aq)$

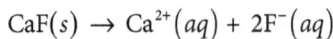

$\left[Ca^{2+}\right] = \left[CaF_2\right] = 1.24\times10^{-3}\ M$

$\left[F^-\right] = 1.24\times10^{-3}\ M \cdot \dfrac{2\ mol\ F^-}{1\ mol\ Ca^{2+}} = 2.48\times10^{-3}\ M$

$K_{sp} = \left[Ca^{2+}\right]\left[F^-\right]^2 = \left(1.24\times10^{-3}\ M\right)\left(2.48\times10^{-3}\ M\right)^2 = 7.63\times10^{-9}$

31.

$$CaSO_4(s) \rightarrow Ca^{2+}(aq) + SO_4^{2-}(aq)$$

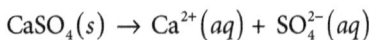

$$0.956 \text{ g } CaSO_4 \cdot \frac{\text{mol}}{136.142 \text{ g}} = 0.007022 \text{ mol}$$

$$\frac{0.007022 \text{ mol}}{1.00 \text{ L}} = 0.007022 \text{ } M$$

$$[Ca^{2+}] = [SO_4^{2-}] = [CaSO_4] = 0.007022 \text{ } M$$

$$K_{sp} = [Ca^{2+}][SO_4^{2-}] = (0.007022 \text{ } M)(0.007022 \text{ } M) = 4.93 \times 10^{-5}$$

32.

$$CuCl(s) \rightarrow Cu^+(aq) + Cl^-(aq)$$

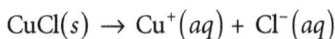

$$\frac{4.11 \times 10^{-3} \text{ g}}{100 \text{ mL}} \cdot \frac{10}{10} = \frac{4.11 \times 10^{-2} \text{ g}}{L}$$

$$\frac{4.11 \times 10^{-2} \text{ g}}{L} \cdot \frac{\text{mol}}{98.999 \text{ g}} = 4.152 \times 10^{-4} \text{ } M \text{ CuCl}$$

$$[Cu^+] = [Cl^-] = [CuCl] = 4.152 \times 10^{-4} \text{ } M$$

$$K_{sp} = [Cu^+][Cl^-] = (4.152 \times 10^{-4})(4.152 \times 10^{-4}) = 1.72 \times 10^{-7}$$

33.

$$PbCl_2(s) \rightarrow Pb^{2+}(aq) + 2Cl^-(aq)$$

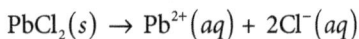

$$\frac{0.45 \text{ g}}{100 \text{ mL}} \cdot \frac{10}{10} = \frac{4.5 \text{ g}}{L}$$

$$\frac{4.5 \text{ g}}{L} \cdot \frac{\text{mol}}{278.1 \text{ g}} = 0.0162 \text{ } M \text{ } PbCl_2$$

$$[Pb^{2+}] = [PbCl_2] = 0.0162 \text{ } M$$

$$[Cl^-] = 0.0162 \text{ } M \cdot \frac{2 \text{ mol } Cl^-}{1 \text{ mol } PbCl_2} = 0.0324 \text{ } M$$

$$K_{sp} = [Pb^{2+}][Cl^-]^2 = (0.0162 \text{ } M)(0.0324 \text{ } M)^2 = 1.7 \times 10^{-5}$$

34.

$$SrF_2(s) \rightarrow Sr^{2+}(aq) + 2F^-(aq)$$

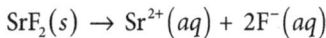

$$\frac{1.29\times10^{-2} \text{ g}}{100 \text{ mL}} \cdot \frac{10}{10} = \frac{0.129 \text{ g}}{L}$$

$$\frac{0.129 \text{ g}}{L} \cdot \frac{\text{mol}}{125.62 \text{ g}} = 1.027\times10^{-3} \text{ } M \text{ SrF}_2$$

$$\left[Sr^{2+}\right] = \left[SrF_2\right] = 1.027\times10^{-3} \text{ } M$$

$$\left[F^-\right] = 1.027\times10^{-3} \text{ } M \cdot \frac{2 \text{ mol F}^-}{1 \text{ mol SrF}_2} = 2.054\times10^{-3} \text{ } M$$

$$K_{sp} = \left[Sr^{2+}\right]\left[F^-\right]^2 = \left(1.027\times10^{-3} \text{ } M\right)\left(2.054\times10^{-3} \text{ } M\right)^2 = 4.33\times10^{-9}$$

35.

$$YF_3(s) \rightarrow Y^{3+}(aq) + 3F^-(aq)$$

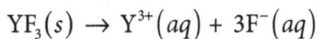

$$K_{sp} = 8.62\times10^{-21}$$

$$K_{sp} = \left[Y^{3+}\right]\left[F^-\right]^3$$

For each mole of YF_3 there is 1 mole of Y^{3+} ions. Thus, $\left[YF_3\right] = \left[Y^{3+}\right]$

For each mole of Y^{3+} ions there are 3 moles of F^- ions. Thus, $\left[F^-\right] = 3\cdot\left[Y^{3+}\right]$.

Let $\left[Y^{3+}\right] = x$. Then $\left[F^-\right] = 3x$.

$$K_{sp} = x(3x)^3 = 27x^4$$

$$27x^4 = 8.62\times10^{-21}$$

$$x^4 = 3.19\times10^{-22}$$

$$x = \left(3.19\times10^{-22}\right)^{\frac{1}{4}} = 4.23\times10^{-6} \text{ } M = \left[YF_3\right]$$

---

36. a.

$$CuI(s) \rightarrow Cu^+(aq) + I^-(aq)$$

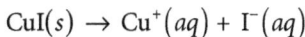

$$K_{sp} = 1.27\times10^{-12}$$

$$K_{sp} = \left[Cu^+\right]\left[I^-\right]$$

For each mole of CuI there is 1 mole of Cu^+ and 1 mole of I^- ions.

Thus, $\left[CuI\right] = \left[Cu^+\right] = \left[I^-\right]$

Let $\left[Cu^+\right] = \left[I^-\right] = x$.

$$K_{sp} = x\cdot x = x^2$$

$$x^2 = 1.27\times10^{-12}$$

$$x = \left(1.27\times10^{-12}\right)^{\frac{1}{2}} = 1.13\times10^{-6} \text{ } M = \left[CuI\right]$$

$$1.13\times10^{-6} \frac{\text{mol}}{L} \cdot \frac{189.6 \text{ g}}{\text{mol}} = 2.14\times10^{-4} \frac{\text{g}}{L} \cdot \frac{0.1}{0.1} = 2.14\times10^{-5} \frac{\text{g}}{100 \text{ mL}}$$

---

36. b.

$PbSO_4(s) \rightarrow Pb^{2+}(aq) + SO_4^{2-}(aq)$

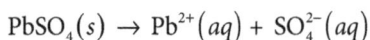

$K_{sp} = 2.53 \times 10^{-8}$

$K_{sp} = \left[Pb^{2+} \right]\left[SO_4^{2-} \right]$

For each mole of $PbSO_4$ there is 1 mole of Pb^{2+} ions and 1 mole of SO_4^{2-} ions.

Thus, $\left[PbSO_4 \right] = \left[Pb^{2+} \right] = \left[SO_4^{2-} \right]$

Let $\left[Pb^{2+} \right] = \left[SO_4^{2-} \right] = x$.

$K_{sp} = x \cdot x = x^2$

$x^2 = 2.53 \times 10^{-8}$

$x = \left(2.53 \times 10^{-8} \right)^{\frac{1}{2}} = 1.59 \times 10^{-4}\ M = \left[PbSO_4 \right]$

$1.59 \times 10^{-4}\ \dfrac{\text{mol}}{\text{L}} \cdot \dfrac{303.3\ \text{g}}{\text{mol}} = 4.82 \times 10^{-2}\ \dfrac{\text{g}}{\text{L}} \cdot \dfrac{0.1}{0.1} = 4.82 \times 10^{-3}\ \dfrac{\text{g}}{100\ \text{mL}}$

---

36. c.

$HgBr(s) \rightarrow Hg^+(aq) + Br^-(aq)$

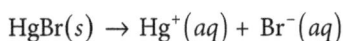

$K_{sp} = 6.40 \times 10^{-23}$

$K_{sp} = \left[Hg^+ \right]\left[Br^- \right]$

For each mole of HgBr there is 1 mole of Hg^+ ions and 1 mole of Br^- ions.

Thus, $\left[HgBr \right] = \left[Hg^+ \right] = \left[Br^- \right]$

Let $\left[Hg^+ \right] = \left[Br^- \right] = x$.

$K_{sp} = x \cdot x = x^2$

$x^2 = 6.40 \times 10^{-23}$

$x = \left(6.40 \times 10^{-23} \right)^{\frac{1}{2}} = 8.00 \times 10^{-12}\ M = \left[HgBr \right]$

$8.00 \times 10^{-12}\ \dfrac{\text{mol}}{\text{L}} \cdot \dfrac{280.5\ \text{g}}{\text{mol}} = 2.24 \times 10^{-9}\ \dfrac{\text{g}}{\text{L}} \cdot \dfrac{0.1}{0.1} = 2.24 \times 10^{-10}\ \dfrac{\text{g}}{100\ \text{mL}}$

---

36. d.

$HgBr_2(s) \rightarrow Hg^{2+}(aq) + 2Br^-(aq)$

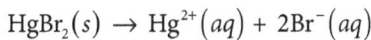

$K_{sp} = 6.20 \times 10^{-20}$

$K_{sp} = \left[Hg^{2+}\right]\left[Br^-\right]^2$

For each mole of $HgBr_2$ there is 1 mole of Hg^+ ions and 2 moles of Br^- ions.

Thus, $\left[HgBr_2\right] = \left[Hg^+\right]$

Let $\left[Hg^+\right] = x$. Then $\left[Br^-\right] = 2x$.

$K_{sp} = x \cdot (2x)^2 = 4x^3$

$4x^3 = 6.20 \times 10^{-20}$

$x^3 = 1.55 \times 10^{-20}$

$x = \left(1.55 \times 10^{-20}\right)^{\frac{1}{3}} = 2.49 \times 10^{-7} \; M = \left[HgBr_2\right]$

$2.49 \times 10^{-7} \; \dfrac{mol}{L} \cdot \dfrac{360.4 \; g}{mol} = 8.99 \times 10^{-5} \; \dfrac{g}{L} \cdot \dfrac{0.1}{0.1} = 8.99 \times 10^{-6} \; \dfrac{g}{100 \; mL}$

---

36. e.

$Cd(OH)_2(s) \rightarrow Cd^{2+}(aq) + 2OH^-(aq)$

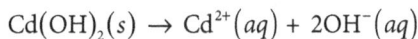

$K_{sp} = 7.2 \times 10^{-15}$

$K_{sp} = \left[Cd^{2+}\right]\left[OH^-\right]^2$

For each mole of $Cd(OH)_2$ there is 1 mole of Cd^{2+} ions and 2 moles of OH^- ions.

Thus, $\left[Cd(OH)_2\right] = \left[Cd^{2+}\right]$

Let $\left[Cd^{2+}\right] = x$. Then $\left[OH^-\right] = 2x$.

$K_{sp} = x \cdot (2x)^2 = 4x^3$

$4x^3 = 7.2 \times 10^{-15}$

$x^3 = 1.8 \times 10^{-15}$

$x = \left(1.8 \times 10^{-15}\right)^{\frac{1}{3}} = 1.22 \times 10^{-5} \; M = \left[Cd(OH)_2\right]$

$1.22 \times 10^{-5} \; \dfrac{mol}{L} \cdot \dfrac{146.4 \; g}{mol} = 1.8 \times 10^{-3} \; \dfrac{g}{L} \cdot \dfrac{0.1}{0.1} = 1.8 \times 10^{-4} \; \dfrac{g}{100 \; mL}$

In this instance K_{sp} only has 2 sig digs, so the result has 2 sig digs.

36. f.

$$Ca_3(PO_4)_2(s) \rightarrow 3Ca^{2+}(aq) + 2PO_4^{3-}(aq)$$

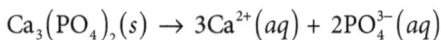

$$K_{sp} = 2.07 \times 10^{-33}$$

$$K_{sp} = \left[Ca^{2+}\right]^3 \left[PO_4^{3-}\right]^2$$

For each mole of $Ca_3(PO_4)_2$ there are 3 moles of Ca^{2+} ions and 2 moles of PO_4^{3-} ions.

Thus, $\left[Ca^{2+}\right] = 3\left[Ca_3(PO_4)_2\right]$

Let $\left[Ca_3(PO_4)_2\right] = x$. Then $\left[Ca^{2+}\right] = 3x$ and $\left[PO_4^{3-}\right] = 2x$.

$$K_{sp} = (3x)^3 \cdot (2x)^2 = 108x^5$$

$$108x^5 = 2.07 \times 10^{-33}$$

$$x^5 = 1.917 \times 10^{-35}$$

$$x = \left(1.917 \times 10^{-35}\right)^{\frac{1}{5}} = 1.139 \times 10^{-7} \; M = \left[Ca_3(PO_4)_2\right]$$

$$1.139 \times 10^{-7} \; \frac{mol}{L} \cdot \frac{310.2 \; g}{mol} = 3.53 \times 10^{-5} \; \frac{g}{L} \cdot \frac{0.1}{0.1} = 3.53 \times 10^{-6} \; \frac{g}{100 \; mL}$$

---

36. g.

$$Tl_2CrO_4(s) \rightarrow 2Tl^+(aq) + CrO_4^{2-}(aq)$$

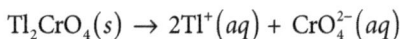

$$K_{sp} = 8.67 \times 10^{-13}$$

$$K_{sp} = \left[Tl^+\right]^2 \left[CrO_4^{2-}\right]$$

For each mole of Tl_2CrO_4 there are 2 moles of Tl^+ ions and 1 mole of CrO_4^{2-} ions.

Thus, $\left[Tl^+\right] = 2\left[Tl_2CrO_4\right]$

Let $\left[Tl_2CrO_4\right] = x$. Then $\left[Tl^+\right] = 2x$ and $\left[CrO_4^{2-}\right] = x$.

$$K_{sp} = (2x)^2 \cdot x = 4x^3$$

$$4x^3 = 8.67 \times 10^{-13}$$

$$x^3 = 2.168 \times 10^{-13}$$

$$x = \left(2.168 \times 10^{-13}\right)^{\frac{1}{3}} = 6.007 \times 10^{-5} \; M = \left[Tl_2CrO_4\right]$$

$$6.007 \times 10^{-5} \; \frac{mol}{L} \cdot \frac{524.8 \; g}{mol} = 3.15 \times 10^{-2} \; \frac{g}{L} \cdot \frac{0.1}{0.1} = 3.15 \times 10^{-3} \; \frac{g}{100 \; mL}$$

---

38. a.

$$Ca(NO_3)_2 + NaOH \rightarrow Ca^{2+} + 2NO_3^- + Na^+ + OH^-$$

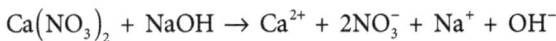

Possible precipitate: $Ca(OH)_2$; $K_{sp} = 5.02 \times 10^{-6}$

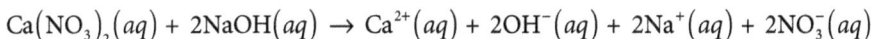

$$Ca(NO_3)_2(aq) + 2NaOH(aq) \rightarrow Ca^{2+}(aq) + 2OH^-(aq) + 2Na^+(aq) + 2NO_3^-(aq)$$

Number of moles of Ca^{2+} = number of moles of $Ca(NO_3)_2$.

$$0.37 \text{ L } Ca(NO_3)_2 \cdot \frac{0.0034 \text{ mol}}{L} = 0.00126 \text{ mol } Ca^{2+}$$

Number of moles of OH^- = number of moles of NaOH.

$$0.25 \text{ L NaOH} \cdot \frac{0.0065 \text{ mol}}{L} = 0.00163 \text{ mol } OH^-$$

Total volume $= 0.37 \text{ L} + 0.25 \text{ L} = 0.62 \text{ L}$

$$\left[Ca^{2+} \right] = \frac{0.00126 \text{ mol}}{0.62 \text{ L}} = 0.00203 \text{ M}$$

$$\left[OH^- \right] = \frac{0.00163}{0.62 \text{ L}} = 0.00263 \text{ M}$$

$$\left[Ca^{2+} \right]\left[OH^- \right]^2 = 0.00203 \text{ M} \cdot (0.00263 \text{ M})^2 = 1.40 \times 10^{-8}$$

Compare to K_{sp}:

$1.40 \times 10^{-8} < 5.02 \times 10^{-6}$. No precipitate forms.

---

38. b.

$$AgNO_3 + Na_2SO_4 \rightarrow Ag^+ + NO_3^- + 2Na^+ + SO_4^{2-}$$

Possible precipitate: Ag_2SO_4; $K_{sp} = 1.20 \times 10^{-5}$

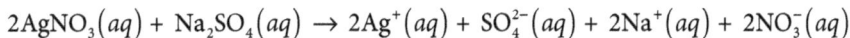

$$2AgNO_3(aq) + Na_2SO_4(aq) \rightarrow 2Ag^+(aq) + SO_4^{2-}(aq) + 2Na^+(aq) + 2NO_3^-(aq)$$

Number of moles of Ag^+ = number of moles of $AgNO_3$.

$$100.0 \text{ mL } AgNO_3 \cdot \frac{1 \text{ L}}{1000 \text{ mL}} \cdot \frac{0.050 \text{ mol}}{L} = 0.0050 \text{ mol } Ag^+$$

Number of moles of SO_4^{2-} = number of moles of Na_2SO_4.

$$10.0 \text{ mL } Na_2SO_4 \cdot \frac{1 \text{ L}}{1000 \text{ mL}} \cdot \frac{0.040 \text{ mol}}{L} = 0.00040 \text{ mol } SO_4^{2-}$$

Total volume $= 100.0 \text{ mL} + 10.0 \text{ mL} = 110.0 \text{ mL} \cdot \frac{1 \text{ L}}{1000 \text{ mL}} = 0.110 \text{ L}$

$$\left[Ag^+ \right] = \frac{0.0050 \text{ mol}}{0.110 \text{ L}} = 0.0455 \text{ M}$$

$$\left[SO_4^{2-} \right] = \frac{0.00040}{0.110 \text{ L}} = 0.00364 \text{ M}$$

$$\left[Ag^+ \right]^2 \left[SO_4^{2-} \right] = (0.0455 \text{ M})^2 \cdot 0.00364 \text{ M} = 7.5 \times 10^{-6}$$

Compare to K_{sp}:

$7.5 \times 10^{-6} < 1.20 \times 10^{-5}$. No precipitate forms.

---

38. c.

$$AgNO_3 + NaCl \rightarrow Ag^+ + NO_3^- + Na^+ + Cl^-$$

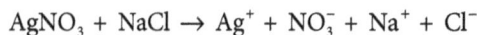

Possible precipitate: AgCl; $K_{sp} = 1.77 \times 10^{-10}$

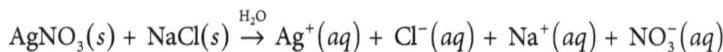

$$AgNO_3(s) + NaCl(s) \xrightarrow{H_2O} Ag^+(aq) + Cl^-(aq) + Na^+(aq) + NO_3^-(aq)$$

Number of moles of Ag^+ = number of moles of $AgNO_3$.

$$1.70 \text{ g AgNO}_3 \cdot \frac{\text{mol}}{169.9 \text{ g}} = 0.0100 \text{ mol Ag}^+$$

Number of moles of Cl^- = number of moles of NaCl.

$$19.5 \text{ g NaCl} \cdot \frac{\text{mol}}{58.44 \text{ g}} = 0.003337 \text{ mol Cl}^-$$

$$\text{Volume} = 175.0 \text{ mL} \cdot \frac{1 \text{ L}}{1000 \text{ mL}} = 0.1750 \text{ L}$$

$$\left[Ag^+ \right] = \frac{0.0100 \text{ mol}}{0.1750 \text{ L}} = 0.05714 \text{ M}$$

$$\left[Cl^- \right] = \frac{0.003337}{0.1750 \text{ L}} = 0.1907 \text{ M}$$

$$\left[Ag^+ \right]\left[Cl^- \right] = 0.05714 \text{ M} \cdot 0.1907 \text{ M} = 0.0109$$

Compare to K_{sp}:

$0.0109 > 1.77 \times 10^{-10}$. Precipitate forms.

---

38. d.

$$Fe(NO_3)_2 + NaOH \rightarrow Fe^{2+} + 2NO_3^- + Na^+ + OH^-$$

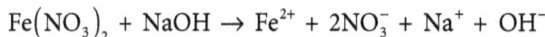

Possible precipitate: $Fe(OH)_2$; $K_{sp} = 4.87 \times 10^{-17}$

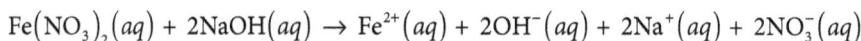

$$Fe(NO_3)_2(aq) + 2NaOH(aq) \rightarrow Fe^{2+}(aq) + 2OH^-(aq) + 2Na^+(aq) + 2NO_3^-(aq)$$

Number of moles of Fe^{2+} = number of moles of $Fe(NO_3)_2$.

$$50.0 \text{ mL Fe}(NO_3)_2 \cdot \frac{1 \text{ L}}{1000 \text{ mL}} \cdot \frac{0.0103 \text{ mol}}{L} = 5.15 \times 10^{-4} \text{ mol Fe}^{2+}$$

Number of moles of OH^- = number of moles of NaOH.

$$45 \text{ mL NaOH} \cdot \frac{1 \text{ L}}{1000 \text{ mL}} \cdot \frac{1.1 \times 10^{-9} \text{ mol}}{L} = 4.95 \times 10^{-11} \text{ mol OH}^-$$

$$\text{Total volume} = 50.0 \text{ mL} + 45 \text{ mL} = 95 \text{ mL} \cdot \frac{1 \text{ L}}{1000 \text{ mL}} = 0.095 \text{ L}$$

$$\left[Fe^{2+} \right] = \frac{5.15 \times 10^{-4} \text{ mol}}{0.095 \text{ L}} = 0.00542 \text{ M}$$

$$\left[OH^- \right] = \frac{4.95 \times 10^{-11}}{0.095 \text{ L}} = 5.21 \times 10^{-10} \text{ M}$$

$$\left[Fe^{2+} \right]\left[OH^- \right]^2 = 0.00542 \text{ M} \cdot \left(5.21 \times 10^{-10} \text{ M}\right)^2 = 1.5 \times 10^{-21}$$

Compare to K_{sp}:

$1.5 \times 10^{-21} < 4.87 \times 10^{-17}$. No precipitate forms.

---

38. e.

$$Na_2CO_3 + BaBr_2 \rightarrow 2Na^+ + CO_3^{2-} + Ba^{2+} + 2Br^-$$

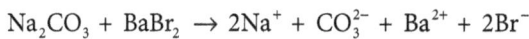

Possible precipitate: $BaCO_3$; $K_{sp} = 2.58 \times 10^{-9}$

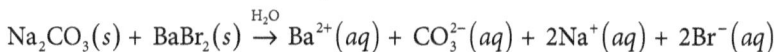

$$Na_2CO_3(s) + BaBr_2(s) \xrightarrow{H_2O} Ba^{2+}(aq) + CO_3^{2-}(aq) + 2Na^+(aq) + 2Br^-(aq)$$

Number of moles of Ba^{2+} = number of moles of $BaBr_2$.

$$0.19 \text{ g BaBr}_2 \cdot \frac{mol}{297.1 \text{ g}} = 6.39 \times 10^{-4} \text{ mol Ba}^{2+}$$

Number of moles of CO_3^{2-} = number of moles of Na_2CO_3.

$$0.97 \text{ g Na}_2CO_3 \cdot \frac{mol}{105.99 \text{ g}} = 0.00915 \text{ mol CO}_3^{2-}$$

Volume = 12.0 L

$$[Ba^{2+}] = \frac{6.39 \times 10^{-4} \text{ mol}}{12.0 \text{ L}} = 5.33 \times 10^{-5} \ M$$

$$[CO_3^{2-}] = \frac{0.00915}{12.0 \text{ L}} = 7.63 \times 10^{-4} \ M$$

$$[Ba^{2+}][CO_3^{2-}] = 5.33 \times 10^{-5} \cdot 7.63 \times 10^{-4} \ M = 4.1 \times 10^{-8}$$

Compare to K_{sp}:

$4.1 \times 10^{-8} > 2.58 \times 10^{-9}$. Precipitate forms.

39.

Assuming a 100-gram sample:

$$64.27 \text{ g C} \cdot \frac{mol}{12.011 \text{ g}} = 5.3509 \text{ mol} \quad (5.3509/1.7838 \approx 3)$$

$$7.19 \text{ g H} \cdot \frac{mol}{1.0079 \text{ g}} = 7.134 \text{ mol} \quad (7.134/1.7838 \approx 4)$$

$$28.54 \text{ g O} \cdot \frac{mol}{15.9994 \text{ g}} = 1.7838 \text{ mol} \quad (1.7838/1.7838 = 1)$$

The ratios give an empirical formula of C_3H_4O. The molecular mass for this empirical formula is 56.064. The given molar mass is three times this amount, so we multiply all subscripts by three to obtain the molecular formula $C_9H_{12}O_3$.

---

40. a.

$$5.00 \text{ mol KMnO}_4 \cdot \frac{5 \text{ mol CO}_2}{14 \text{ mol KMnO}_4} = 1.79 \text{ mol CO}_2$$

---

40. b.

$$2.66 \text{ kg} \cdot \frac{1000 \text{ g}}{1 \text{ kg}} \cdot \frac{\text{mol}}{92.09 \text{ g}} = 28.88 \text{ mol } C_3H_5(OH)_3$$

$$28.88 \text{ mol } C_3H_5(OH)_3 \cdot \frac{16 \text{ mol } H_2O}{4 \text{ mol } C_3H_5(OH)_3} = 115.5 \text{ mol } H_2O$$

$$115.5 \text{ mol } H_2O \cdot \frac{18.02 \text{ g}}{\text{mol}} = 2080 \text{ g} \cdot \frac{1 \text{ kg}}{1000 \text{ g}} = 2.08 \text{ kg}$$

---

40. c.

$$500.0 \text{ g} \cdot \frac{1000 \text{ g}}{1 \text{ kg}} \cdot \frac{\text{mol}}{92.09 \text{ g}} = 5.429 \text{ mol } C_3H_5(OH)_3$$

$$5.429 \text{ mol } C_3H_5(OH)_3 \cdot \frac{14 \text{ mol } KMnO_4}{4 \text{ mol } C_3H_5(OH)_3} = 19.00 \text{ mol } KMnO_4$$

$$19.00 \text{ mol } KMnO_4 \cdot \frac{158.03 \text{ g}}{\text{mol}} = 3003 \text{ g}$$

---

41. a.

$$4.36 \text{ mol } CH_3CH_2OH \cdot \frac{1 \text{ mol } CH_3COOCH_2CH_3}{1 \text{ mol } CH_3CH_2OH} = 4.36 \text{ mol } CH_3COOCH_2CH_3$$

$$\frac{2.97 \text{ mol } CH_3COOCH_2CH_3}{4.36 \text{ mol } CH_3COOCH_2CH_3} \cdot 100\% = 68.1\%$$

---

41. b.

Since the reactant mole ratio is 1:1, ethanol is the limiting reactant. The mole ratio of ethanol to ethyl acetate is also 1:1, so

$$\frac{2.30 \text{ mol } CH_3COOCH_2CH_3}{2.90 \text{ mol } CH_3COOCH_2CH_3} \cdot 100\% = 79.3\%$$

42.

$V_1 = 4.99 \text{ L}$

$P_1 = 2.15 \text{ atm}$

$$P_2 = 1.00 \text{ bar} \cdot \frac{100,000 \text{ Pa}}{1 \text{ bar}} \cdot \frac{1 \text{ atm}}{101,325 \text{ Pa}} = 0.9869 \text{ atm}$$

$$P_1V_1 = P_2V_2 \rightarrow V_2 = V_1 \cdot \frac{P_1}{P_2} = 4.99 \text{ L} \cdot \frac{2.15 \text{ atm}}{0.9869 \text{ atm}} = 10.9 \text{ L}$$

---

43. a.

Vapor pressure of water at these conditions: 3.7831 kPa

$P_T = 205.15 \text{ kPa}$

$P_T = P_{H_2S} + P_{H_2O}$

$P_{H_2S} = P_T - P_{H_2O} = 205.15 \text{ kPa} - 3.7831 \text{ kPa} = 201.37 \text{ kPa}$

---

43. b.

$P_{H_2S} = 201.37$ kPa

$V = 15.3 \text{ mL} \cdot \dfrac{1 \text{ L}}{1000 \text{ mL}} = 0.0153 \text{ L}$

$T = 28°C \rightarrow 28°C + 273.2 = 301.2 \text{ K}$

$R = 8.314 \dfrac{\text{L} \cdot \text{kPa}}{\text{mol} \cdot \text{K}}$

$PV = nRT \rightarrow n = \dfrac{PV}{RT} = \dfrac{201.37 \text{ kPa} \cdot 0.0153 \text{ L}}{8.314 \dfrac{\text{L} \cdot \text{kPa}}{\text{mol} \cdot \text{K}} \cdot 301.2 \text{ K}} = 0.00123 \text{ mol}$

---

43. c

$0.00123 \text{ mol} \cdot \dfrac{34.08 \text{ g}}{\text{mol}} = 0.0419 \text{ g}$

44.

$2N_2O_5 \rightarrow 4NO_2 + O_2$

$100.0 \text{ g} \cdot \dfrac{\text{mol}}{108.01 \text{ g}} = 0.92584 \text{ mol N}_2O_5$

$0.92584 \text{ mol N}_2O_5 \cdot \dfrac{1 \text{ mol O}_2}{2 \text{ mol N}_2O_5} = 0.46292 \text{ mol O}_2$

$n = 0.46292 \text{ mol}$

$P = 100 \text{ kPa}$

$T = 273.15 \text{ K}$

$R = 8.314 \dfrac{\text{L} \cdot \text{kPa}}{\text{mol} \cdot \text{K}}$

$PV = nRT \rightarrow V = \dfrac{nRT}{P} = \dfrac{0.46292 \text{ mol} \cdot 8.314 \dfrac{\text{L} \cdot \text{kPa}}{\text{mol} \cdot \text{K}} \cdot 273.15 \text{ K}}{100 \text{ kPa}} = 10.51 \text{ L}$

45.

$K_f = 1.86 \dfrac{°C}{m}$

$n = 5.62 \text{ g} \cdot \dfrac{\text{mol}}{110.983 \text{ g}} = 0.05064 \text{ mol}$

$V = 100.0 \text{ mL} \cdot \dfrac{1 \text{ L}}{1000 \text{ mL}} = 0.1000 \text{ L}$

$m = \dfrac{0.05064 \text{ mol}}{0.1000 \text{ L}} = 0.5064 \ m$

$\Delta T_f = K_f m = 1.86 \dfrac{°C}{m} \cdot 0.5064 \ m = 0.942°C$

$T_f = 0.00°C - 0.942°C = -0.94°C$

47.

$2NaOH + 2Na \rightarrow 2Na_2O + H_2$

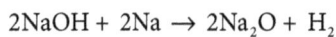

	Reactants			Products		
	2NaOH	2Na	Sum (Σ)	$2Na_2O$	H_2	Sum (Σ)
ΔG_f° kJ/mol	2(−397.7)		−795.4	2(−375.5)		−751.0

$$\Delta G° = \Sigma\Delta G°_{f,\text{products}} - \Sigma\Delta G°_{f,\text{reactants}} = -751.0 \ \frac{kJ}{mol} - \left(-795.4 \ \frac{kJ}{mol}\right) = 44.4 \ \frac{kJ}{mol}$$

Chapter 12

6. a.

Add one H_2O to the left to balance the oxygen. Add $2H^+$ to the right to balance the hydrogen. Add $2OH^-$ to each side to "neutralize" the $2H^+$. Add two electrons to the right to balance the charges.

$$SO_3^{2-}(aq) + H_2O(l) \rightarrow SO_4^{2-}(aq) + 2H^+(aq)$$
$$SO_3^{2-}(aq) + H_2O(l) + 2OH^-(aq) \rightarrow SO_4^{2-}(aq) + 2H^+(aq) + 2OH^-(aq)$$
$$SO_3^{2-}(aq) + H_2O(l) + 2OH^-(aq) \rightarrow SO_4^{2-}(aq) + 2H^+(aq) + 2OH^-(aq) + 2e^-$$

On the right, combine the H^+ and OH^- ions to form $2H_2O$. Cancel out one H_2O from each side.

$$SO_3^{2-}(aq) + H_2O(l) + 2OH^-(aq) \rightarrow SO_4^{2-}(aq) + 2H_2O(l) + 2e^-$$
$$SO_3^{2-}(aq) + 2OH^-(aq) \rightarrow SO_4^{2-}(aq) + H_2O(l) + 2e^-$$

6. b.

Begin by placing a coefficient of 2 on the NH_3 to balance the nitrogen. Add $6H^+$ to the left to balance the hydrogen. Add $6OH^-$ to each side to neutralize the H^+. Combine the ions on the left to form $6H_2O$. Add six electrons to the left to balance the charges.

$$N_2(g) \rightarrow 2NH_3(g)$$
$$N_2(g) + 6H^+(aq) + 6OH^-(aq) \rightarrow 2NH_3(g) + 6OH^-(aq)$$
$$N_2(g) + 6H_2O(l) \rightarrow 2NH_3(g) + 6OH^-(aq)$$
$$N_2(g) + 6H_2O(l) + 6e^- \rightarrow 2NH_3(g) + 6OH^-(aq)$$

6. c.

$$N_2(g) \rightarrow 2NH_4^+(aq)$$
$$N_2(g) + 8H^+(aq) \rightarrow 2NH_4^+(aq)$$
$$N_2(g) + 8H^+(aq) + 6e^- \rightarrow 2NH_4^+(aq)$$

6. d.

The only thing needed is two electrons on the right to balance the charges.

$$Sn^{2+}(aq) \rightarrow Sn^{4+}(aq) + 2e^-$$

6. e.

$$ClO_3^-(aq) + 6H^+(aq) \rightarrow Cl^-(aq) + 3H_2O(l)$$
$$ClO_3^-(aq) + 6H^+(aq) + 6e^- \rightarrow Cl^-(aq) + 3H_2O(l)$$

6. f.

$$OH^-(aq) + H_2O(l) \rightarrow O_2(g) + 3H^+(aq)$$
$$OH^-(aq) + H_2O(l) + 3OH^-(aq) \rightarrow O_2(g) + 3H^+(aq) + 3OH^-(aq)$$
$$H_2O(l) + 4OH^-(aq) \rightarrow O_2(g) + 3H_2O(l)$$
$$4OH^-(aq) \rightarrow O_2(g) + 2H_2O(l)$$
$$4OH^-(aq) \rightarrow O_2(g) + 2H_2O(l) + 4e^-$$

6. g.

$$H_2SO_3(aq) + H_2O(l) \rightarrow SO_4^{2-}(aq) + 4H^+(aq)$$
$$H_2SO_3(aq) + H_2O(l) \rightarrow SO_4^{2-}(aq) + 4H^+(aq) + 2e^-$$

6. h.

$$O_2(g) + 4H^+(aq) \rightarrow H_2O(l) + H_2O(l)$$
$$O_2(g) + 4H^+(aq) + 4OH^-(aq) \rightarrow 2H_2O(l) + 4OH^-(aq)$$
$$O_2(g) + 4H_2O \rightarrow 2H_2O(l) + 4OH^-(aq)$$
$$O_2(g) + 2H_2O + 4e^- \rightarrow 4OH^-(aq)$$

7.

In the following solutions, the first set of equations isolates the species involved in the oxidation and reduction. The second set of equations derives the complete pair of half-reactions. The third set of equations shows the addition of the half-reactions, cancellations, and addition of ions to form complete compounds.

7. a.

$$FeCl_3(aq) + H_2S(aq) \rightarrow FeCl_2(aq) + HCl(aq) + S(s)$$
$$Fe^{3+}(aq) + 3Cl^-(aq) + 2H^+(aq) + S^{2-}(aq) \rightarrow Fe^{2+}(aq) + 2Cl^-(aq) + H^+(aq) + Cl^-(aq) + S(s)$$
$$Fe^{3+}(aq) + S^{2-}(aq) \rightarrow Fe^{2+}(aq) + S(s)$$

$$2Fe^{3+}(aq) + 2e^- \rightarrow 2Fe^{2+}(aq)$$
$$S^{2-}(aq) \rightarrow S(s) + 2e$$

$$2Fe^{3+}(aq) + S^{2-}(aq) \rightarrow 2Fe^{2+}(aq) + S(s)$$
$$2Fe^{3+}(aq) + S^{2-}(aq) + 6Cl^-(aq) + 2H^+(aq) \rightarrow 2Fe^{2+}(aq) + S(s) + 6Cl^-(aq) + 2H^+(aq)$$
$$2FeCl_3(aq) + H_2S(aq) \rightarrow 2FeCl_2(aq) + S(s) + 2HCl(aq)$$

7. b.

$$I_2(s) + HClO(aq) \rightarrow HIO_3(aq) + HCl(aq)$$
$$I_2(s) + H^+(aq) + ClO^-(aq) \rightarrow H^+(aq) + IO_3^-(aq) + H^+(aq) + Cl^-(aq)$$
$$I_2(s) + ClO^-(aq) \rightarrow IO_3^-(aq) + Cl^-(aq)$$

$$I_2(s) \rightarrow IO_3^-(aq)$$

$$I_2(s) \rightarrow 2IO_3^-(aq)$$

$$I_2(s) + 6H_2O(l) \rightarrow 2IO_3^-(aq) + 12H^+(aq)$$

$$I_2(s) + 6H_2O(l) \rightarrow 2IO_3^-(aq) + 12H^+(aq) + 10e^-$$

$$ClO^-(aq) \rightarrow Cl^-(aq)$$

$$ClO^-(aq) + 2H^+(aq) \rightarrow Cl^-(aq) + H_2O(l)$$

$$ClO^-(aq) + 2H^+(aq) + 2e^- \rightarrow Cl^-(aq) + H_2O(l)$$

$$I_2(s) + 6H_2O(l) \rightarrow 2IO_3^-(aq) + 12H^+(aq) + 10e^-$$

$$5ClO^-(aq) + 10H^+(aq) + 10e^- \rightarrow 5Cl^-(aq) + 5H_2O(l)$$

$$I_2(s) + 6H_2O(l) + 5ClO^-(aq) + 10H^+(aq) \rightarrow 2IO_3^-(aq) + 12H^+(aq) + 5Cl^-(aq) + 5H_2O(l)$$

$$I_2(s) + H_2O(l) + 5HClO(aq) + 5H^+(aq) \rightarrow 2HIO_3(aq) + 5H^+(aq) + 5HCl(aq)$$

$$I_2(s) + H_2O(l) + 5HClO(aq) \rightarrow 2HIO_3(aq) + 5HCl(aq)$$

---

7. c.

$$KMnO_4(aq) + CH_3OH(aq) \rightarrow Mn(OH)_2(aq) + HCOOH(aq) + KOH(aq)$$

$$K^+(aq) + MnO_4^-(aq) + CH_3OH(aq) \rightarrow$$
$$Mn^{2+}(aq) + 2OH^-(aq) + H^+(aq) + COOH^-(aq) + K^+(aq) + OH^-(aq)$$

$$MnO_4^-(aq) + CH_3OH(aq) \rightarrow Mn^{2+}(aq) + COOH^-(aq)$$

$$MnO_4^-(aq) \rightarrow Mn^{2+}(aq)$$

$$MnO_4^-(aq) + 8H^+(aq) \rightarrow Mn^{2+}(aq) + 4H_2O(l)$$

$$MnO_4^-(aq) + 8H^+(aq) + 5e^- \rightarrow Mn^{2+}(aq) + 4H_2O(l)$$

$$CH_3OH(aq) \rightarrow COOH^-(aq)$$

$$CH_3OH(aq) + H_2O(l) \rightarrow COOH^-(aq) + 5H^+(aq)$$

$$CH_3OH(aq) + H_2O(l) \rightarrow COOH^-(aq) + 5H^+(aq) + 4e^-$$

$$4MnO_4^-(aq) + 32H^+(aq) + 20e^- \rightarrow 4Mn^{2+}(aq) + 16H_2O(l)$$

$$5CH_3OH(aq) + 5H_2O(l) \rightarrow 5COOH^-(aq) + 25H^+(aq) + 20e^-$$

$$4MnO_4^-(aq) + 32H^+(aq) + 5CH_3OH(aq) + 5H_2O(l) \rightarrow$$
$$4Mn^{2+}(aq) + 16H_2O(l) + 5COOH^-(aq) + 25H^+(aq)$$

$$4MnO_4^-(aq) + 12H^+(aq) + 5CH_3OH(aq) \rightarrow 4Mn^{2+}(aq) + 11H_2O(l) + 5HCOOH(aq)$$

$$4MnO_4^-(aq) + 12H^+(aq) + 5CH_3OH(aq) + 4K^+(aq) + 12OH^-(aq) \rightarrow$$
$$4Mn^{2+}(aq) + 11H_2O(l) + 5HCOOH(aq) + 4K^+(aq) + 12OH^-(aq)$$

$$4KMnO_4(aq) + 5CH_3OH(aq) + 12H_2O \rightarrow$$
$$4Mn(OH)_2(aq) + 11H_2O(l) + 5HCOOH(aq) + 4KOH(aq)$$

$$4KMnO_4(aq) + 5CH_3OH(aq) + H_2O \rightarrow 4Mn(OH)_2(aq) + 5HCOOH(aq) + 4KOH(aq)$$

---

7. d.

$$H_2O_2(aq) + ClO_2(aq) \rightarrow HClO_2(aq) + O_2(g)$$
$$H_2O_2(aq) + ClO_2(aq) \rightarrow H^+(aq) + ClO_2^-(aq) + O_2(g)$$
$$H_2O_2(aq) + ClO_2(aq) \rightarrow ClO_2^-(aq) + O_2(g)$$

$$H_2O_2(aq) \rightarrow O_2(g)$$
$$H_2O_2(aq) + 2OH^-(aq) \rightarrow O_2(g) + 2H^+(aq) + 2OH^-(aq)$$
$$H_2O_2(aq) + 2OH^-(aq) \rightarrow O_2(g) + 2H^+(aq) + 2OH^-(aq) + 2e^-$$
$$ClO_2(aq) \rightarrow ClO_2^-(aq)$$
$$ClO_2(aq) + e^- \rightarrow ClO_2^-(aq)$$

$$H_2O_2(aq) + 2OH^-(aq) \rightarrow O_2(g) + 2H^+(aq) + 2OH^-(aq) + 2e^-$$
$$2ClO_2(aq) + 2e^- \rightarrow 2ClO_2^-(aq)$$
$$H_2O_2(aq) + 2OH^-(aq) + 2ClO_2(aq) \rightarrow O_2(g) + 2H^+(aq) + 2OH^-(aq) + 2ClO_2^-(aq)$$
$$H_2O_2(aq) + 2OH^-(aq) + 2ClO_2(aq) \rightarrow O_2(g) + 2H_2O(l) + 2ClO_2^-(aq)$$
$$H_2O_2(aq) + 2OH^-(aq) + 2ClO_2(aq) + 2H^+(aq) \rightarrow O_2(g) + 2H_2O(l) + 2ClO_2^-(aq) + 2H^+(aq)$$
$$H_2O_2(aq) + 2H_2O(l) + 2ClO_2(aq) \rightarrow O_2(g) + 2H_2O(l) + 2HClO_2(aq)$$
$$H_2O_2(aq) + 2ClO_2(aq) \rightarrow O_2(g) + 2HClO_2(aq)$$

---

7. e.

$$Al(s) + LiCl(aq) + LiNO_2(aq) \rightarrow LiAlO_2(aq) + NH_4Cl(aq)$$
$$Al(s) + Li^+(aq) + Cl^-(aq) + Li^+(aq) + NO_2^-(aq) \rightarrow Li^+(aq) + AlO_2^-(aq) + NH_4^+(aq) + Cl^-(aq)$$
$$Al(s) + NO_2^-(aq) \rightarrow AlO_2^-(aq) + NH_4^+(aq)$$

$$Al(s) \rightarrow AlO_2^-(aq)$$
$$Al(s) + 2H_2O(l) \rightarrow AlO_2^-(aq) + 4H^+(aq)$$
$$Al(s) + 2H_2O(l) + 4OH^-(aq) \rightarrow AlO_2^-(aq) + 4H^+(aq) + 4OH^-(aq)$$
$$Al(s) + 2H_2O(l) + 4OH^-(aq) \rightarrow AlO_2^-(aq) + 4H^+(aq) + 4OH^-(aq) + 3e^-$$
$$NO_2^-(aq) \rightarrow NH_4^+(aq)$$
$$NO_2^-(aq) + 8H^+(aq) \rightarrow NH_4^+(aq) + 2H_2O(l)$$
$$NO_2^-(aq) + 8H^+(aq) + 8OH^-(aq) \rightarrow NH_4^+(aq) + 2H_2O(l) + 8OH^-(aq)$$
$$NO_2^-(aq) + 8H^+(aq) + 8OH^-(aq) + 6e^- \rightarrow NH_4^+(aq) + 2H_2O(l) + 8OH^-(aq)$$

$$2Al(s) + 4H_2O(l) + 8OH^-(aq) \rightarrow 2AlO_2^-(aq) + 8H^+(aq) + 8OH^-(aq) + 6e^-$$
$$NO_2^-(aq) + 8H^+(aq) + 8OH^-(aq) + 6e^- \rightarrow NH_4^+(aq) + 2H_2O(l) + 8OH^-(aq)$$
$$2Al(s) + 4H_2O(l) + 8OH^-(aq) + NO_2^-(aq) + 8H^+(aq) + 8OH^-(aq) \rightarrow$$
$$2AlO_2^-(aq) + 8H^+(aq) + 8OH^-(aq) + NH_4^+(aq) + 2H_2O(l) + 8OH^-(aq)$$
$$2Al(s) + 12H_2O(l) + 8OH^-(aq) + NO_2^-(aq) \rightarrow$$
$$2AlO_2^-(aq) + NH_4^+(aq) + 10H_2O(l) + 8OH^-(aq)$$
$$2Al(s) + 2H_2O(l) + 8OH^-(aq) + NO_2^-(aq) \rightarrow$$
$$2AlO_2^-(aq) + NH_4^+(aq) + 8OH^-(aq)$$
$$2Al(s) + 2H_2O(l) + 8OH^-(aq) + NO_2^-(aq) + 2Li^+(aq) + Cl^-(aq) \rightarrow$$
$$2AlO_2^-(aq) + NH_4^+(aq) + 8OH^-(aq) + 2Li^+(aq) + Cl^-(aq)$$
$$2Al(s) + 2H_2O(l) + LiNO_2(aq) + LiCl(aq) \rightarrow 2LiAlO_2(aq) + NH_4Cl(aq)$$

---

7. f.

$$HCl(aq) + HMnO_4(aq) \rightarrow Cl_2(g) + MnCl_2(aq)$$
$$H^+(aq) + Cl^-(aq) + H^+(aq) + MnO_4^-(aq) \rightarrow Cl_2(g) + Mn^{2+}(aq) + 2Cl^-(aq)$$
$$Cl^-(aq) + MnO_4^-(aq) \rightarrow Cl_2(g) + Mn^{2+}(aq) + 2Cl^-(aq)$$

$$Cl^-(aq) \rightarrow Cl_2(g) + 2Cl^-(aq)$$
$$4Cl^-(aq) \rightarrow Cl_2(g) + 2Cl^-(aq) + 2e^-$$
$$MnO_4^-(aq) \rightarrow Mn^{2+}(aq)$$
$$MnO_4^-(aq) + 8H^+(aq) \rightarrow Mn^{2+}(aq) + 4H_2O(l)$$
$$MnO_4^-(aq) + 8H^+(aq) + 5e^- \rightarrow Mn^{2+}(aq) + 4H_2O(l)$$

$$20Cl^-(aq) \rightarrow 5Cl_2(g) + 10Cl^-(aq) + 10e^-$$
$$2MnO_4^-(aq) + 16H^+(aq) + 10e^- \rightarrow 2Mn^{2+}(aq) + 8H_2O(l)$$
$$20Cl^-(aq) + 2MnO_4^-(aq) + 16H^+(aq) \rightarrow 5Cl_2(g) + 10Cl^-(aq) + 2Mn^{2+}(aq) + 8H_2O(l)$$
$$14HCl(aq) + 6Cl^-(aq) + 2HMnO_4(aq) \rightarrow 5Cl_2(g) + 6Cl^-(aq) + 2MnCl_2(aq) + 8H_2O(l)$$
$$14HCl(aq) + 2HMnO_4(aq) \rightarrow 5Cl_2(g) + 2MnCl_2(aq) + 8H_2O(l)$$

---

7. g.

$$S(s) + HNO_3(aq) \rightarrow H_2SO_3(aq) + N_2O(g)$$
$$S(s) + H^+(aq) + NO_3^-(aq) \rightarrow 2H^+(aq) + SO_3^{2-}(aq) + N_2O(g)$$
$$S(s) + NO_3^-(aq) \rightarrow SO_3^{2-}(aq) + N_2O(g)$$

$$S(s) \rightarrow SO_3^{2-}(aq)$$
$$S(s) + 3H_2O(l) \rightarrow SO_3^{2-}(aq) + 6H^+(aq)$$
$$S(s) + 3H_2O(l) \rightarrow SO_3^{2-}(aq) + 6H^+(aq) + 4e^-$$
$$NO_3^-(aq) \rightarrow N_2O(g)$$
$$2NO_3^-(aq) \rightarrow N_2O(g)$$
$$2NO_3^-(aq) + 10H^+(aq) \rightarrow N_2O(g) + 5H_2O(l)$$
$$2NO_3^-(aq) + 10H^+(aq) + 8e^- \rightarrow N_2O(g) + 5H_2O(l)$$

$$2S(s) + 6H_2O(l) \rightarrow 2SO_3^{2-}(aq) + 12H^+(aq) + 8e^-$$
$$2NO_3^-(aq) + 10H^+(aq) + 8e^- \rightarrow N_2O(g) + 5H_2O(l)$$
$$2S(s) + 6H_2O(l) + 2NO_3^-(aq) + 10H^+(aq) \rightarrow 2SO_3^{2-}(aq) + 12H^+(aq) + N_2O(g) + 5H_2O(l)$$
$$2S(s) + H_2O(l) + 2HNO_3(aq) + 8H^+(aq) \rightarrow 2H_2SO_3(aq) + 8H^+(aq) + N_2O(g)$$
$$2S(s) + H_2O(l) + 2HNO_3(aq) \rightarrow 2H_2SO_3(aq) + N_2O(g)$$

---

7. h.

$$HBr(aq) + HMnO_4(aq) \rightarrow HBrO_3(aq) + MnO_2(s)$$
$$H^+(aq) + Br^-(aq) + H^+(aq) + MnO_4^-(aq) \rightarrow H^+(aq) + BrO_3^-(aq) + MnO_2(s)$$
$$Br^-(aq) + MnO_4^-(aq) \rightarrow BrO_3^-(aq) + MnO_2(s)$$

$$Br^-(aq) \rightarrow BrO_3^-(aq)$$
$$Br^-(aq) + 3H_2O(l) \rightarrow BrO_3^-(aq) + 6H^+(aq)$$
$$Br^-(aq) + 3H_2O(l) \rightarrow BrO_3^-(aq) + 6H^+(aq) + 6e^-$$
$$MnO_4^-(aq) \rightarrow MnO_2(s)$$
$$MnO_4^-(aq) + 4H^+(aq) \rightarrow MnO_2(s) + 2H_2O(l)$$
$$MnO_4^-(aq) + 4H^+(aq) + 3e^- \rightarrow MnO_2(s) + 2H_2O(l)$$

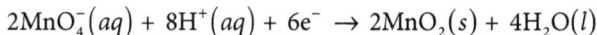

$$Br^-(aq) + 3H_2O(l) \rightarrow BrO_3^-(aq) + 6H^+(aq) + 6e^-$$
$$2MnO_4^-(aq) + 8H^+(aq) + 6e^- \rightarrow 2MnO_2(s) + 4H_2O(l)$$
$$HBr(aq) + 2HMnO_4(aq) + 5H^+(aq) \rightarrow HBrO_3(aq) + 5H^+(aq) + 2MnO_2(s) + H_2O(l)$$
$$HBr(aq) + 2HMnO_4(aq) \rightarrow HBrO_3(aq) + 2MnO_2(s) + H_2O(l)$$

9. d.

$$65 \text{ g Fe}_2O_3 \cdot \frac{\text{mol}}{159.69 \text{ g}} = 0.407 \text{ mol Fe}_2O_3$$

$$0.407 \text{ mol Fe}_2O_3 \cdot \frac{2 \text{ mol Al}}{1 \text{ mol Fe}_2O_3} = 0.814 \text{ mol Al}$$

$$0.814 \text{ mol Al} \cdot \frac{26.98 \text{ g}}{\text{mol}} = 22 \text{ g Al}$$

9. e.

$$65 \text{ g Fe}_2O_3 \cdot \frac{\text{mol}}{159.69 \text{ g}} = 0.407 \text{ mol Fe}_2O_3$$

$$0.407 \text{ mol Fe}_2O_3 \cdot \frac{2 \text{ mol Fe}}{1 \text{ mol Fe}_2O_3} = 0.814 \text{ mol Fe}$$

$$0.814 \text{ mol Fe} \cdot \frac{55.85 \text{ g}}{\text{mol}} = 45 \text{ g Fe}$$

19. a.

$$E^\circ_{cell} = E_{cathode} - E_{anode} = 0.7996 \text{ V} - (-0.1375 \text{ V}) = 0.9371 \text{ V}$$

19. b.

$$E^\circ_{cell} = E_{cathode} - E_{anode} = 0.771 \text{ V} - (-0.28 \text{ V}) = 1.05 \text{ V}$$

19. c.

$$E^\circ_{cell} = E_{cathode} - E_{anode} = 1.066 \text{ V} - (-2.71 \text{ V}) = 3.78 \text{ V}$$

19. d.

$$E^\circ_{cell} = E_{cathode} - E_{anode} = 1.066 \text{ V} - (0.5355 \text{ V}) = 0.530 \text{ V}$$

24. a.

$$1.27 \text{ g NO}_2 \cdot \frac{\text{mol}}{46.01 \text{ g}} = 0.02761 \text{ mol NO}_2$$

$$0.02761 \text{ mol NO}_2 \cdot \frac{1 \text{ mol N}_2\text{O}_5}{2 \text{ mol NO}_2} = 0.01380 \text{ mol N}_2\text{O}_5$$

$$0.01380 \text{ mol N}_2\text{O}_5 \cdot \frac{108.01 \text{ g}}{\text{mol}} = 1.49 \text{ g}$$

$$\frac{1.20 \text{ g}}{1.49 \text{ g}} \cdot 100\% = 80.5\%$$

---

24. b.

$$4.0 \text{ mol O}_3 \cdot \frac{1 \text{ mol O}_2}{1 \text{ mol O}_3} = 4.0 \text{ mol O}_2$$

$$4.0 \text{ mol O}_2 \cdot \frac{32.00 \text{ g}}{\text{mol}} = 128 \text{ g}$$

$$\frac{x}{128 \text{ g}} = 0.760 \rightarrow x = 0.760 \cdot 128 \text{ g} = 97 \text{ g}$$

---

24. c.

$$4.304 \text{ g NO}_2 \cdot \frac{\text{mol}}{46.01 \text{ g}} = 0.09354 \text{ mol NO}_2$$

$$0.09354 \text{ mol NO}_2 \cdot \frac{1 \text{ mol O}_3}{2 \text{ mol NO}_2} = 0.04677 \text{ mol O}_3$$

$$0.04677 \text{ mol O}_3 \cdot \frac{48.00 \text{ g}}{\text{mol}} = 2.245 \text{ g O}_3$$

More than this much O_3 is available, so NO_2 is the limiting reactant.

25.

$P_1 = 62.4$ atm, gauge

$T_1 = 22.00°\text{C} \rightarrow 22.00°\text{C} + 273.15 = 295.15 \text{ K}$

$T_2 = 35.00°\text{C} \rightarrow 35.00°\text{C} + 273.15 = 308.15 \text{ K}$

$P_{1,\text{abs}} = P_{\text{gauge}} + P_{\text{atm}} = 62.4 \text{ atm} + 1.0 \text{ atm} = 63.4 \text{ atm}$

$$PV = nRT \rightarrow \frac{P}{T} = \frac{nR}{V} = \text{constant}$$

$$\frac{P_1}{T_1} = \frac{P_2}{T_2} \rightarrow P_2 = P_1 \cdot \frac{T_2}{T_1} = 63.4 \text{ atm} \cdot \frac{308.15 \text{ K}}{295.15 \text{ K}} = 66.2 \text{ atm}$$

$P_{2,\text{gauge}} = P_{2,\text{abs}} - P_{\text{atm}} = 66.2 \text{ atm} - 1.0 \text{ atm} = 65.2 \text{ atm, gauge}$

26.

$T = 16.0°C \rightarrow 16.0°C + 273.15 = 289.15 \text{ K}$

$P_T = 102,360 \text{ Pa} \cdot \dfrac{1 \text{ kPa}}{100,000 \text{ Pa}} = 102.360 \text{ kPa}$

$V = 275 \text{ mL} \cdot \dfrac{1 \text{ L}}{1000 \text{ mL}} = 0.275 \text{ L}$

At these conditions, the vapor pressure of water is 2.0647 kPa

$P_T = P_{\text{ethane}} + P_{\text{water}}$

$P_{\text{ethane}} = P_T - P_{\text{water}} = 102.360 \text{ kPa} - 2.0647 \text{ kPa} = 100.295 \text{ kPa}$

$R = 8.314 \dfrac{\text{L} \cdot \text{kPa}}{\text{mol} \cdot \text{K}}$

$PV = nRT \rightarrow n = \dfrac{PV}{RT} = \dfrac{100.295 \text{ kPa} \cdot 0.275 \text{ L}}{8.314 \dfrac{\text{L} \cdot \text{kPa}}{\text{mol} \cdot \text{K}} \cdot 289.15 \text{ K}} = 0.0115 \text{ mol}$

27.

$200 \cdot 325 \text{ mg} \cdot \dfrac{1 \text{ g}}{1000 \text{ mg}} = 65.0 \text{ g } C_9H_8O_4$

$65.0 \text{ g } C_9H_8O_4 \cdot \dfrac{\text{mol}}{180.2 \text{ g}} = 0.3608 \text{ mol } C_9H_8O_4$

$0.3608 \text{ mol } C_9H_8O_4 \cdot \dfrac{6.022 \times 10^{23} \text{ particles}}{\text{mol}} = 2.173 \times 10^{23} \text{ particles}$

Each particle is a molecule containing 9 carbon atoms.

$2.173 \times 10^{23} \cdot 9 = 1.96 \times 10^{24} \text{ C atoms}$

28.

$CH_3CH_2OH + 3O_2 \rightarrow 3H_2O + 2CO_2$

$1.00 \text{ L } H_2O \cdot \dfrac{1000 \text{ mL}}{1 \text{ L}} \cdot \dfrac{998 \text{ g}}{\text{mL}} = 998 \text{ g } H_2O$

$998 \text{ g } H_2O \cdot \dfrac{\text{mol}}{18.02 \text{ g}} = 55.38 \text{ mol } H_2O$

$55.38 \text{ mol } H_2O \cdot \dfrac{1 \text{ mol } CH_3CH_2OH}{3 \text{ mol } H_2O} = 18.46 \text{ mol } CH_3CH_2OH$

$18.46 \text{ mol } CH_3CH_2OH \cdot \dfrac{46.07 \text{ g}}{\text{mol}} = 850.5 \text{ g } CH_3CH_2OH$

$850.5 \text{ g } CH_3CH_2OH \cdot \dfrac{\text{cm}^3}{0.789 \text{ g}} = 1080 \text{ cm}^3 = 1080 \text{ mL} \cdot \dfrac{1 \text{ L}}{1000 \text{ mL}} = 1.08 \text{ L}$

29.

$$C_2H_6 + \frac{7}{2}O_2 \rightarrow 2CO_2 + 3H_2O$$

	Reactants			Products		
	C_2H_6	$(7/2)O_2$	Sum (Σ)	$2CO_2$	$3H_2O$	Sum (Σ)
ΔG_f° kJ/mol	−32.0		−32.0	2(−394.4)	3(−237.1)	−1500.1

$$\Delta G^\circ = \Sigma \Delta G^\circ_{f,\,products} - \Sigma \Delta G^\circ_{f,\,reactants} = -1500.1 \ \frac{kJ}{mol} - \left(-32.0 \ \frac{kJ}{mol}\right) = -1468.1 \ \frac{kJ}{mol}$$

Chapter 13

41. a.

$$2C_8H_{18} + 25O_2 \rightarrow 16CO_2 + 18H_2O$$

41. b.

$$1.00 \text{ L} \cdot \frac{703 \text{ g}}{\text{L}} = 703 \text{ g } C_8H_{18}$$

$$703 \text{ g } C_8H_{18} \cdot \frac{\text{mol}}{114.2 \text{ g}} = 6.154 \text{ mol } C_8H_{18}$$

$$6.154 \text{ mol } C_8H_{18} \cdot \frac{25 \text{ mol } O_2}{2 \text{ mol } C_8H_{18}} = 76.93 \text{ mol } O_2$$

Since air is 20.95% O_2, 0.2095 of total moles of air particles are O_2.

$$0.2095 \cdot n_T = 76.93 \text{ mol} \rightarrow n_T = \frac{76.93 \text{ mol}}{0.2095} = 367.2 \text{ mol}$$

$P = 100 \text{ kPa}$

$T = 273.15 \text{ K}$

$$R = 8.314 \ \frac{\text{L} \cdot \text{kPa}}{\text{mol} \cdot \text{K}}$$

$$PV = nRT \rightarrow V = \frac{nRT}{P} = \frac{367.2 \text{ mol} \cdot 8.314 \ \dfrac{\text{L} \cdot \text{kPa}}{\text{mol} \cdot \text{K}} \cdot 273.15 \text{ K}}{100 \text{ kPa}} = 8340 \text{ L}$$

41. c.

$$1.00 \text{ L} \cdot \frac{703 \text{ g}}{\text{L}} = 703 \text{ g } C_8H_{18}$$

$$703 \text{ g } C_8H_{18} \cdot \frac{\text{mol}}{114.2 \text{ g}} = 6.154 \text{ mol } C_8H_{18}$$

$$6.154 \text{ mol } C_8H_{18} \cdot \frac{18 \text{ mol } H_2O}{2 \text{ mol } C_8H_{18}} = 55.39 \text{ mol } H_2O$$

$$55.39 \text{ mol } H_2O \cdot \frac{18.02 \text{ g}}{\text{mol}} = 998 \text{ g}$$

42.

$V_1 = 8340 \text{ L}$

$T_1 = 273.15 \text{ K}$

$T_2 = 546°C \rightarrow 546°C + 273.2°C = 819.2 \text{ K}$

$$\frac{V_1}{T_1} = \frac{V_2}{T_2} \rightarrow V_2 = V_1 \cdot \frac{T_2}{T_1} = 8340 \text{ L} \cdot \frac{819.2 \text{ K}}{273.15 \text{ K}} = 25,010 \text{ L}$$

Writing this value with 3 sig digs requires scientific notation, giving 2.50×10^4 L.

43. a.

$$625 \text{ g CH}_3\text{COOH} \cdot \frac{\text{mol}}{60.05 \text{ g}} = 10.41 \text{ mol CH}_3\text{COOH}$$

$$M = \frac{10.41 \text{ mol}}{1.00 \text{ L}} = 10.4 \ M$$

43. b.

$$10.4 \ \frac{\text{mol}}{\text{L}} \cdot 0.013 = 0.135 \ M = \left[\text{H}_3\text{O}^+\right]$$

$$\text{pH} = -\log\left[\text{H}_3\text{O}^+\right] = -\log 0.135 = 0.870$$

$$\text{pH} + \text{pOH} = 14.0$$

$$\text{pOH} = 14.0 - \text{pH} = 14.0 - 0.870 = 13.1$$

$$\left[\text{H}_3\text{O}^+\right]\left[\text{OH}^-\right] = 1.0 \times 10^{-14} \ M^2$$

$$\left[\text{OH}^-\right] = \frac{1.0 \times 10^{-14} \ M^2}{\left[\text{H}_3\text{O}^+\right]} = \frac{1.0 \times 10^{-14} \ M^2}{0.135 \ M} = 7.4 \times 10^{-14} \ M$$

43. c.

$$0.135 \ \frac{\text{mol}}{\text{L}} \cdot 1.00 \text{ L} = 0.135 \text{ mol H}_3\text{O}^+ \rightarrow 0.135 \text{ mol OH}^- \rightarrow 0.135 \text{ mol NaOH}$$

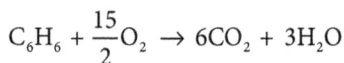

$$0.135 \text{ mol NaOH} \cdot \frac{\text{L}}{1.00 \text{ mol}} = 0.135 \text{ L NaOH}$$

44. a.

$$2\text{C}_6\text{H}_6 + 15\text{O}_2 \rightarrow 12\text{CO}_2 + 6\text{H}_2\text{O}$$

44. b.

$$\text{C}_6\text{H}_6 + \frac{15}{2}\text{O}_2 \rightarrow 6\text{CO}_2 + 3\text{H}_2\text{O}$$

	Reactants			Products		
	C_6H_6	15O_2	Sum (Σ)	6CO_2	$3\text{H}_2\text{O}$	Sum (Σ)
ΔH_f° kJ/mol	49.1	0	49.1	6(−393.5)	3(−285.8)	−3218.4

$$\Delta H^\circ = \Sigma \Delta H^\circ_{f,\text{products}} - \Sigma \Delta H^\circ_{f,\text{reactants}} = -3218.4 \ \frac{\text{kJ}}{\text{mol}} - 49.1 \ \frac{\text{kJ}}{\text{mol}} = -3267.5 \ \frac{\text{kJ}}{\text{mol}}$$

45.

$$CH_3CH_2OH + 3O_2 \rightarrow 3H_2O + 2CO_2$$

	Reactants			Products		
	CH_3CH_2OH	$3O_2$	Sum (Σ)	$2CO_2$	$3H_2O$	Sum (Σ)
ΔG_f° kJ/mol	-174.8		-174.8	$2(-394.4)$	$3(-237.1)$	-1500.1

$$\Delta G^\circ = \Sigma \Delta G^\circ_{f,\,products} - \Sigma \Delta G^\circ_{f,\,reactants} = -1500.1\ \frac{kJ}{mol} - \left(-174.8\ \frac{kJ}{mol}\right) = -1325.3\ \frac{kJ}{mol}$$

46.

$pH = 2.60$

$\left[H_3O^+\right] = 10^{-2.60} = 0.0025\ M$

$\left[H_3O^+\right] = \left[HCOO^-\right]$

$[HCOOH] = 0.035\ M$

$$K_a = \frac{\left[H_3O^+\right]\left[HCOO^-\right]}{[HCOOH]} = \frac{0.0025 \cdot 0.0025}{0.035} = 1.8 \times 10^{-4}$$

47.

$$CaCl_2 + 2NaOH \rightarrow 2NaCl + Ca(OH)_2$$

Possible precipitate: $Ca(OH)_2$

$Ca(OH)_2 \quad Ca^{2+} + 2OH^-$

$K_{sp} = 5.02 \times 10^{-6}$

$$645\ mL \cdot \frac{1\ L}{1000\ mL} = 0.645\ L\ NaOH$$

$$0.645\ L\ NaOH \cdot \frac{0.024\ mol}{L} = 0.0155\ mol\ NaOH \rightarrow 0.0155\ mol\ OH^-$$

$$300.0\ mL \cdot \frac{1\ L}{1000\ mL} = 0.3000\ L\ CaCl_2$$

$$0.3000\ L\ CaCl_2 \cdot \frac{0.111\ mol}{L} = 0.0333\ mol\ CaCl_2 \rightarrow 0.0333\ mol\ Ca^{2+}$$

Total volume $= 0.645\ L + 0.300\ L = 0.945\ L$

$$\left[Ca^{2+}\right] = \frac{0.0333\ mol}{0.945\ L} = 0.0352\ M$$

$$\left[OH^-\right] = \frac{0.0155\ mol}{0.945\ L} = 0.0160\ M$$

$$\left[Ca^{2+}\right]\left[OH^-\right]^2 = 0.0352\ M \cdot (0.0160\ M)^2 = 9.01 \times 10^{-6}$$

$9.01 \times 10^{-6} > 5.02 \times 10^{-6}$, a precipitate will form

48. a.

$$CuS(s) + HNO_3(aq) \rightarrow Cu(NO_3)_2(aq) + NO(g) + S(s) + H_2O(l)$$
$$CuS(s) + H^+(aq) + NO_3^-(aq) \rightarrow Cu^{2+}(aq) + 2NO_3^-(aq) + NO(g) + S(s) + H_2O(l)$$
$$CuS(s) + NO_3^-(aq) \rightarrow Cu^{2+}(aq) + 2NO_3^-(aq) + NO(g) + S(s)$$

$$CuS(s) \rightarrow Cu^{2+}(aq) + S(s)$$
$$CuS(s) \rightarrow Cu^{2+}(aq) + S(s) + 2e^-$$
$$NO_3^-(aq) \rightarrow 2NO_3^-(aq) + NO(g)$$
$$3NO_3^-(aq) \rightarrow 2NO_3^-(aq) + NO(g)$$
$$3NO_3^-(aq) + 4H^+(aq) \rightarrow 2NO_3^-(aq) + NO(g) + 2H_2O(l)$$
$$3NO_3^-(aq) + 4H^+(aq) + 3e^- \rightarrow 2NO_3^-(aq) + NO(g) + 2H_2O(l)$$

$$3CuS(s) \rightarrow 3Cu^{2+}(aq) + 3S(s) + 6e^-$$
$$6NO_3^-(aq) + 8H^+(aq) + 6e^- \rightarrow 4NO_3^-(aq) + 2NO(g) + 4H_2O(l)$$
$$3CuS(s) + 6NO_3^-(aq) + 8H^+(aq) \rightarrow 3Cu^{2+}(aq) + 3S(s) + 4NO_3^-(aq) + 2NO(g) + 4H_2O(l)$$
$$3CuS(s) + 6NO_3^-(aq) + 8H^+(aq) + 2NO_3^-(aq) \rightarrow$$
$$3Cu^{2+}(aq) + 3S(s) + 4NO_3^-(aq) + 2NO(g) + 4H_2O(l) + 2NO_3^-(aq)$$
$$3CuS(s) + 8HNO_3(aq) \rightarrow 3Cu(NO_2)_2(aq) + 3S(s) + 2NO(g) + 4H_2O(l)$$

---

48. b.

$$Fe(s) + SnCl_4(aq) \rightarrow FeCl_3(aq) + SnCl_2(aq)$$
$$Fe(s) + Sn^{4+}(aq) + 4Cl^-(aq) \rightarrow Fe^{3+}(aq) + 3Cl^-(aq) + Sn^{2+}(aq) + 2Cl^-(aq)$$
$$Fe(s) + Sn^{4+}(aq) \rightarrow Fe^{3+}(aq) + Sn^{2+}(aq)$$

$$Fe(s) \rightarrow Fe^{3+}(aq)$$
$$Fe(s) \rightarrow Fe^{3+}(aq) + 3e^-$$
$$Sn^{4+}(aq) \rightarrow Sn^{2+}(aq)$$
$$Sn^{4+}(aq) + 2e^- \rightarrow Sn^{2+}(aq)$$

$$2Fe(s) \rightarrow 2Fe^{3+}(aq) + 6e^-$$
$$3Sn^{4+}(aq) + 6e^- \rightarrow 3Sn^{2+}(aq)$$
$$2Fe(s) + 3Sn^{4+}(aq) \rightarrow 2Fe^{3+}(aq) + 3Sn^{2+}(aq)$$
$$2Fe(s) + 3Sn^{4+}(aq) + 12Cl^-(aq) \rightarrow 2Fe^{3+}(aq) + 3Sn^{2+}(aq) + 12Cl^-(aq)$$
$$2Fe(s) + 3SnCl_4(aq) \rightarrow 2FeCl_3(aq) + 3SnCl_2(aq)$$

---

49. a.

$$Bi(OH)_3(s) + K_2SnO_2(aq) \rightarrow Bi(s) + K_2SnO_3(aq)$$
$$Bi(OH)_3(s) + 2K^-(aq) + SnO_2^{2-}(aq) \rightarrow Bi(s) + 2K^-(aq) + SnO_3^{2-}(aq)$$
$$Bi(OH)_3(s) + SnO_2^{2-}(aq) \rightarrow Bi(s) + SnO_3^{2-}(aq)$$

$$Bi(OH)_3(s) \rightarrow Bi(s)$$
$$Bi(OH)_3(s) + 3H^+(aq) \rightarrow Bi(s) + 3H_2O(l)$$
$$Bi(OH)_3(s) + 3H^+(aq) + 3OH^-(aq) \rightarrow Bi(s) + 3H_2O(l) + 3OH^-(aq)$$
$$Bi(OH)_3(s) + 3H^+(aq) + 3OH^-(aq) + 3e^- \rightarrow Bi(s) + 3H_2O(l) + 3OH^-(aq)$$
$$SnO_2^{2-}(aq) \rightarrow SnO_3^{2-}(aq)$$
$$SnO_2^{2-}(aq) + H_2O(l) \rightarrow SnO_3^{2-}(aq) + 2H^+(aq)$$
$$SnO_2^{2-}(aq) + H_2O(l) + 2OH^-(aq) \rightarrow SnO_3^{2-}(aq) + 2H^+(aq) + 2OH^-(aq)$$
$$SnO_2^{2-}(aq) + H_2O(l) + 2OH^-(aq) \rightarrow SnO_3^{2-}(aq) + 2H^+(aq) + 2OH^-(aq) + 2e^-$$

$$2Bi(OH)_3(s) + 6H^+(aq) + 6OH^-(aq) + 6e^- \rightarrow 2Bi(s) + 6H_2O(l) + 6OH^-(aq)$$
$$3SnO_2^{2-}(aq) + 3H_2O(l) + 6OH^-(aq) \rightarrow 3SnO_3^{2-}(aq) + 6H^+(aq) + 6OH^-(aq) + 6e^-$$
$$2Bi(OH)_3(s) + 6H^+(aq) + 6OH^-(aq) + 3SnO_2^{2-}(aq) + 3H_2O(l) + 6OH^-(aq) \rightarrow$$
$$2Bi(s) + 6H_2O(l) + 6OH^-(aq) + 3SnO_3^{2-}(aq) + 6H^+(aq) + 6OH^-(aq)$$
$$2Bi(OH)_3(s) + 3SnO_2^{2-}(aq) + 9H_2O(l) + 6OH^-(aq) \rightarrow$$
$$2Bi(s) + 12H_2O(l) + 6OH^-(aq) + 3SnO_3^{2-}(aq)$$
$$2Bi(OH)_3(s) + 3SnO_2^{2-}(aq) + 6OH^-(aq) \rightarrow$$
$$2Bi(s) + 3H_2O(l) + 6OH^-(aq) + 3SnO_3^{2-}(aq)$$
$$2Bi(OH)_3(s) + 3SnO_2^{2-}(aq) + 6K^+(aq) \rightarrow$$
$$2Bi(s) + 3H_2O(l) + 3SnO_3^{2-}(aq) + 6K^+(aq)$$
$$2Bi(OH)_3(s) + 3K_2SnO_2(aq) \rightarrow 2Bi(s) + 3H_2O(l) + 3K_2SnO_3(aq)$$

---

49. b.

$$CO_2(g) + NH_2OH(aq) \rightarrow N_2(g) + CO(g) + H_2O(l)$$
$$CO_2(g) + NH_2^+(aq) + OH^-(aq) \rightarrow N_2(g) + CO(g) + H_2O(l)$$
$$CO_2(g) + NH_2^+(aq) \rightarrow N_2(g) + CO(g)$$

$$CO_2(g) \rightarrow CO(g)$$

$$CO_2(g) + 2H^+(aq) \rightarrow CO(g) + H_2O(l)$$

$$CO_2(g) + 2H^+(aq) + 2OH^-(aq) \rightarrow CO(g) + H_2O(l) + 2OH^-(aq)$$

$$CO_2(g) + 2H^+(aq) + 2OH^-(aq) + 2e^- \rightarrow CO(g) + H_2O(l) + 2OH^-(aq)$$

$$NH_2^+(aq) \rightarrow N_2(g)$$

$$2NH_2^+(aq) \rightarrow N_2(g)$$

$$2NH_2^+(aq) \rightarrow N_2(g) + 4H^+(aq)$$

$$2NH_2^+(aq) + 4OH^-(aq) \rightarrow N_2(g) + 4H^+(aq) + 4OH^-(aq)$$

$$2NH_2^+(aq) + 4OH^-(aq) \rightarrow N_2(g) + 4H^+(aq) + 4OH^-(aq) + 2e^-$$

$$CO_2(g) + 2H^+(aq) + 2OH^-(aq) + 2e^- \rightarrow CO(g) + H_2O(l) + 2OH^-(aq)$$

$$2NH_2^+(aq) + 4OH^-(aq) \rightarrow N_2(g) + 4H^+(aq) + 4OH^-(aq) + 2e^-$$

$$CO_2(g) + 2H_2O(l) + 2NH_2^+(aq) + 4OH^-(aq) \rightarrow CO(g) + H_2O(l) + 2OH^-(aq) + N_2(g) + 4H_2O(l)$$

$$CO_2(g) + 2NH_2OH(aq) + 2OH^-(aq) \rightarrow CO(g) + 2OH^-(aq) + N_2(g) + 3H_2O(l)$$

$$CO_2(g) + 2NH_2OH(aq) \rightarrow CO(g) + N_2(g) + 3H_2O(l)$$

Appendix A

1.

$$1750 \text{ m} \cdot \frac{1 \text{ ft}}{0.3048 \text{ m}} = 5740 \text{ ft}$$

2.

$$3.54 \text{ g} \cdot \frac{1 \text{ kg}}{1000 \text{ g}} = 0.00354 \text{ kg}$$

3.

$$41.11 \text{ mL} \cdot \frac{1 \text{ L}}{1000 \text{ mL}} = 0.04111 \text{ L}$$

4.

$$7 \times 10^8 \text{ m} \cdot \frac{1 \text{ mi}}{1609 \text{ m}} = 400,000 \text{ mi} = 4 \times 10^5 \text{ mi}$$

5.

$$1.5499 \times 10^{-12} \text{ mm} \cdot \frac{1 \text{ m}}{1000 \text{ mm}} = 1.5499 \times 10^{-15} \text{ m}$$

6.

$$750 \text{ cm}^3 \cdot \frac{1 \text{ m}}{100 \text{ cm}} \cdot \frac{1 \text{ m}}{100 \text{ cm}} \cdot \frac{1 \text{ m}}{100 \text{ cm}} = 0.00075 \text{ m}^3 = 7.5 \times 10^{-4} \text{ m}^3$$

7.

$$2.9979 \times 10^8 \ \frac{\text{m}}{\text{s}} \cdot \frac{1 \text{ ft}}{0.3048 \text{ m}} = 9.836 \times 10^8 \ \frac{\text{ft}}{\text{s}}$$

8.

$$168 \text{ hr} \cdot \frac{3600 \text{ s}}{1 \text{ hr}} = 605,000 \text{ s} = 6.05 \times 10^5 \text{ s}$$

If the problem had said exactly one week, we would calculate 604,800 s. But we were given a number of hours with 3 sig digs, so that is what we have in the result.

9.

$$5570 \ \frac{\text{kg}}{\text{m}^3} \cdot \frac{1000 \text{ g}}{1 \text{ kg}} \cdot \frac{1 \text{ m}}{100 \text{ cm}} \cdot \frac{1 \text{ m}}{100 \text{ cm}} \cdot \frac{1 \text{ m}}{100 \text{ cm}} = 5.57 \ \frac{\text{g}}{\text{cm}^3}$$

10.

$$45 \, \frac{\text{gal}}{\text{s}} \cdot \frac{3.785 \, \text{L}}{1 \, \text{gal}} \cdot \frac{1 \, \text{m}^3}{1000 \, \text{L}} \cdot \frac{60 \, \text{s}}{1 \, \text{min}} = 10 \, \frac{\text{m}^3}{\text{min}}$$

To express this result with the required 2 sig digs, we must write it as $1.0 \times 10^1 \, \text{m}^3/\text{min}$.

11.

$$600,000 \, \frac{\text{ft}^3}{\text{s}} \cdot \frac{12 \, \text{in}}{1 \, \text{ft}} \cdot \frac{12 \, \text{in}}{1 \, \text{ft}} \cdot \frac{12 \, \text{in}}{1 \, \text{ft}} \cdot \frac{2.54 \, \text{cm}}{1 \, \text{in}} \cdot \frac{2.54 \, \text{cm}}{1 \, \text{in}} \cdot \frac{2.54 \, \text{cm}}{1 \, \text{in}} \cdot \frac{1 \, \text{L}}{1000 \, \text{cm}^3} \cdot \frac{3600 \, \text{s}}{1 \, \text{hr}} = 6 \times 10^{10} \, \frac{\text{L}}{\text{hr}}$$

12.

$$5200 \, \text{mL} \cdot \frac{1 \, \text{cm}^3}{1 \, \text{mL}} \cdot \frac{1 \, \text{m}}{100 \, \text{cm}} \cdot \frac{1 \, \text{m}}{100 \, \text{cm}} \cdot \frac{1 \, \text{m}}{100 \, \text{cm}} = 0.0052 \, \text{m}^3 = 5.2 \times 10^{-3} \, \text{m}^3$$

13.

$$5.65 \times 10^2 \, \text{mm}^2 \cdot \frac{1 \, \text{m}}{1000 \, \text{mm}} \cdot \frac{1 \, \text{m}}{1000 \, \text{mm}} \cdot \frac{100 \, \text{cm}}{1 \, \text{m}} \cdot \frac{100 \, \text{cm}}{1 \, \text{m}} \cdot \frac{1 \, \text{in}}{2.54 \, \text{cm}} \cdot \frac{1 \, \text{in}}{2.54 \, \text{cm}} = 0.876 \, \text{in}^2$$

14.

$$32.16 \, \frac{\text{ft}}{\text{s}^2} \cdot \frac{0.3048 \, \text{m}}{1 \, \text{ft}} = 9.802 \, \frac{\text{m}}{\text{s}^2}$$

15.

$$5001 \, \frac{\mu\text{g}}{\text{s}} \cdot \frac{1 \, \text{g}}{1 \times 10^6 \, \mu\text{g}} \cdot \frac{1 \, \text{kg}}{1000 \, \text{g}} \cdot \frac{60 \, \text{s}}{1 \, \text{min}} = 3.001 \times 10^{-4} \, \frac{\text{kg}}{\text{min}}$$

16.

$$4.771 \, \frac{\text{g}}{\text{mL}} \cdot \frac{1 \, \text{kg}}{1000 \, \text{g}} \cdot \frac{1000 \, \text{mL}}{1 \, \text{L}} \cdot \frac{1000 \, \text{L}}{1 \, \text{m}^3} = 4771 \, \frac{\text{kg}}{\text{m}^3}$$

17.

$$13.6 \, \frac{\text{g}}{\text{cm}^3} \cdot \frac{1000 \, \text{mg}}{1 \, \text{g}} \cdot \frac{100 \, \text{cm}}{1 \, \text{m}} \cdot \frac{100 \, \text{cm}}{1 \, \text{m}} \cdot \frac{100 \, \text{cm}}{1 \, \text{m}} = 1.36 \times 10^{10} \, \frac{\text{mg}}{\text{m}^3}$$

18.

$$93,000,000 \, \text{mi} \cdot \frac{1609 \, \text{m}}{1 \, \text{mi}} \cdot \frac{100 \, \text{cm}}{1 \, \text{m}} = 1.5 \times 10^{13} \, \text{cm}$$

19.

$$65 \, \frac{\text{mi}}{\text{hr}} \cdot \frac{1609 \, \text{m}}{1 \, \text{mi}} \cdot \frac{1 \, \text{hr}}{3600 \, \text{s}} = 29 \, \frac{\text{m}}{\text{s}}$$

20.

$$633 \text{ nm} \cdot \frac{1 \text{ m}}{10^9 \text{ nm}} \cdot \frac{100 \text{ cm}}{1 \text{ m}} \cdot \frac{1 \text{ in}}{2.54 \text{ cm}} = 0.0000249 \text{ in} = 2.49 \times 10^{-5} \text{ in}$$

21.

First convert the speed of light to mi/hr:

$$2.9979 \times 10^8 \frac{\text{m}}{\text{s}} \cdot \frac{1 \text{ mi}}{1609 \text{ m}} \cdot \frac{3600 \text{ s}}{1 \text{ hr}} = 6.708 \times 10^8 \frac{\text{mi}}{\text{hr}}$$

Now calculate the proportion indicated by the percentage:

$$6.708 \times 10^8 \frac{\text{mi}}{\text{hr}} \cdot 0.05015 = 3.364 \times 10^7 \frac{\text{mi}}{\text{hr}} = 33,640,000 \frac{\text{mi}}{\text{hr}}$$

www.ingramcontent.com/pod-product-compliance
Lightning Source LLC
Chambersburg PA
CBHW081543220326
41598CB00036B/6543